MATH IN ASTRONOMY

By
ROBERT SADLER, Ph.D.

COPYRIGHT © 1997 Mark Twain Media, Inc.

ISBN 1-58037-038-1

Printing No. CD-1883

Mark Twain Media, Inc., Publishers
Distributed by Carson-Dellosa Publishing Company, Inc.

The purchase of this book entitles the buyer to reproduce the student pages for classroom use only. Other permissions may be obtained by writing Mark Twain Media, Inc., Publishers.

All rights reserved. Printed in the United States of America.

Table of Contents

Introduction: To the Teacher 1

Lesson One: Introduction to Astronomy ... 2
   Questions 5
   Math 6
      Numbers in Scientific Notation 6
      Distances 8
      Angles 10
   Teacher's Page 11

Lesson Two: The Solar System 12
   Questions 17
   Math 19
      Time 19
      The Astronomical Unit 20
      Kepler's Laws 22
      Scaling 28
   Teacher's Page 29

Lesson Three: Mercury and Venus—The Inferior Planets 30
   Questions 36
   Math 37
      Mercury and Venusian Years 37
      Force 38
      Mass 39
      Weight 40
      Weight and Mass 41
      Volume 42
      Density 43
      Temperature 44
   Teacher's Page 46

Lesson Four: The Earth 47
   Questions 51
   Math 53
      Hours and Minutes 53
      The Twenty-Four-Hour Clock 55
      Adding and Subtracting Times 56
      Sidereal and Solar Days 58
      Latitude and Longitude 60
      Elevation of the Sun 62
   Teacher's Page 64

Lesson Five: The Moon 65
   Questions 70
   Math 72
      Finding Averages 72
      Using Scaling to Find Sizes 74
      Sizes of Features on the Moon 75
   Teacher's Page 77

Lesson Six: Mars 78
   Questions 81
   Math 82
      Travel Times to the Planets 82
      Communication Times to the Planets 84
   Teacher's Page 86

Lesson Seven: The Outer Planets 87
   Questions 92
   Math 94
      The Mass of Jupiter and Other Planets ... 94
   Teacher's Page 96

Lesson Eight: Asteroids, Comets, and Meteors 97
   Questions 101
   Math 103
      Sequences 103
      The Titius-Bode Rule 104
   Teacher's Page 105

Lesson Nine: The Sun and the Stars 106
   Questions 112
   Math 115
      Distances in Light Years 115
      The Inverse Square Law of Radiation ... 117
   Teacher's Page 118

Lesson Ten: The Milky Way and Other Galaxies 119
   Questions 122
   Math 123
      Hubble's Law 123
      The Redshift 124
      The Age of the Universe 125
   Teacher's Page 126

Introduction: To the Teacher

Astronomy is thought to be the oldest of the sciences, and it has a rich history that extends back to the earliest recorded civilizations. Records from early cultures in Egypt and the Middle East indicate that people were using the heavenly bodies to keep track of and to predict seasonal events. There is evidence that the massive stone structure at Stonehenge in England and even the Big Horn Medicine Wheel in the American West may have been aligned to highlight the summer solstice, an astronomical event. There is also some indication—markings on 27,000-year-old mammoth tusks—that even the very early, primitive peoples observed and made note of the sky.

The early Greeks were the first to attempt to understand the heavens from a rational point of view. Thales of Miletus (624–547 B.C.) was perhaps the first to propose that the universe was rational and understandable to humans, and Pythagoras (570–500 B.C.) suggested that the universe might be mathematical. Many other Greek philosophers contributed to the understanding of the cosmos, and Aristotle (384–322 B.C.) made the Greek view popular with his teachings. Unfortunately, Aristotle's teachings were held in such high regard that his geocentric model, which placed the earth at the center of the universe, was not abandoned until the mid 1500s.

The advances in our understanding of the universe and our place in the cosmos have paralleled discoveries in mathematics. Nicholas Copernicus (1473–1543) used mathematical arguments as well as other evidence to assert that the Sun was the center of the solar system.

Johannes Kepler's (1571–1630) laws of planetary motion, Isaac Newton's (1642–1727) laws of motion and gravitation, and Albert Einstein's (1879–1955) theories of Special and General Relativity are just three examples of astronomical theories and laws that involve an understanding of advanced mathematics.

The twentieth century, particularly the post World War II era, has seen an explosion in the accumulation of astronomical data and the development of astronomical models. Some of this knowledge has come from space vehicles that have orbited Earth, or orbited, flew by, or landed on our Moon or other planets. Most of the new information has been incorporated into quantitative models that are expressed in mathematical form.

In its most pristine form, astronomy is a quantitative science that requires a considerable background in mathematics. In reality, however, many people enjoy astronomy without delving into the quantitative complexities. Some are stargazers who simply observe and appreciate the Moon, planets, and stars and the often spectacular events that can be seen in the sky. Others are historians who have noted the relation of astronomical events and advances in astronomy to the time line of history. Still others are interested in other branches of science such as physics, chemistry, geology, meteorology, or even biology, and have found that astronomy offers applications or examples from their area of study. Finally, many astronomers are amateurs. Unlike other scientific areas, astronomy offers the same laboratory, the bodies in the night sky, to professional and amateur alike. It is not uncommon for amateurs to make modest, but important, contributions to the field.

Astronomy has many faces, and it can be approached in many ways. However, as we will see in this text, even a small amount of mathematics can significantly enhance a person's understanding.

Lesson One
Introduction to Astronomy

Janet had gone to a vacation cabin in the mountains with her parents and her older sister for two weeks. The cabin was a long way from the nearest town and was located next to a mountain lake. Janet enjoyed her stay a lot. The days were filled with hiking, swimming, and fishing, and the evenings were fun because her parents let her stay up much later than usual.

In the late evenings, she began to notice the sky in a way that she had never seen it before in the city. The stars seemed very bright and there appeared to be more of them than she remembered seeing at home. One of them seemed especially bright. Also, there was a milky-gray patch of light that seemed to stretch all of the way across the sky. It was so dim that she had never seen it in the city, but out in the mountains it was visible every night.

One night, as she was sitting with her sister on the boat dock at the lake, Janet began to ask about the sky. How many stars were there in the sky? How far away were they? What was the very bright star—was it a star or something else? Why did the Moon seem to have a different shape on different nights? And what was that milky-gray patch that extended across the sky? It seemed that, once she got started, she would never run out of questions. Her sister answered some of them, but she didn't have all of the answers. The study of the sky, she said, was sometimes called astronomy and she was not an astronomer. Janet decided to find out more about astronomy.

What is Astronomy?

Astronomy is sometimes defined as the study of the heavenly bodies such as the Moon and the stars. A definition such as this, however, makes astronomy sound like a difficult subject that only a professional scientist could enjoy. Nothing could be further from the truth.

Each of us can appreciate the night sky and the objects there in our own way. Scientists study the stars to find out what they are and what causes them to shine. Sailors, on the other hand, observe the positions of the stars to help them navigate on the seas far from land. Artists have painted the starry sky in many ways, and young couples often think that it is romantic. Other people excitedly follow the adventures of the starship *Enterprise* and its crew in *Star Trek* episodes on television and in the movies. There are only about 10,000 professional astronomers in the whole world, but many, many more people are interested in the night sky.

Math in Astronomy

Lesson One: Introduction to Astronomy

What is There to Learn About Astronomy?

What can ordinary people learn about astronomy and the night sky? Many things. First, we live on a spherical body called a planet. We call our planet Earth and feel that it is a special place, but in some ways it is not really special at all. A **planet** is just a heavenly body that travels around the Sun in a path that is almost, but not quite, a circle. The almost circular path that a planet follows is called its **orbit**. Our Earth is in the company of eight other planets that also move in orbits around the Sun. Some of them are larger than Earth, and a few of them are smaller. Like Earth, however, they are spherical in shape. Each of their orbits is of a different size so they do not run into each other.

The earth is also circled by a small body that we call the Moon. It is closer to us than any other heavenly body, and it travels in an orbit around the earth instead of around the Sun. Like Earth and the other planets, the Moon is shaped like a sphere. Because it is so close, we can see some of the patterns on its surface. As we will see later, other planets also have moons, in some cases many moons.

The Earth

When we think about astronomy and observing the sky, we usually think about the Moon, the stars, and sometimes the planets. There is one astronomical body, however, that is visible in the daytime—the Sun. The Sun is by far the largest heavenly body near the earth. It is so large that even the largest planets are tiny by comparison. The Sun generates its own light, and we depend on that light for many things. Sunlight not only brightens our world, but also causes plants to grow and drives the weather processes that bring us winds, clouds, rain, and snow. Without the Sun, life on Earth would not be possible. Also, without the Sun, we would not be able to see the Moon and other planets. The Moon and planets do not create their own light. We see them because of the sunlight that reflects off of their surfaces.

The tiny points of light that we see in the night sky are **stars**. Like the Sun, they shine with their own light. In fact, the Sun is just a star that appears big and bright to us because it is so close to us. The other stars are so far away that they appear very small, like bright little dots scattered across the dark sky.

The stars are not scattered evenly throughout space. Instead, they are clustered into groups called **galaxies**. A galaxy can contain billions of stars. Our Sun is one of the many stars in a galaxy called the **Milky Way**. We see the Milky Way at night as a milky-gray path across the sky. It gets its color from the vast number of stars that it contains.

A Galaxy

The galaxies are almost unbelievably large and are separated by huge distances. There are very few stars and not a lot of anything else in the space between them.

Is there something that is bigger than a galaxy—maybe something that contains galaxies? The answer is yes. We call it the **universe**. All of the stars, all of the galaxies, in fact everything, belongs to the universe.

Patterns in the Stars

In the modern world, most people live in towns or cities where bright street lights and commercial illumination light up the sky at night and make it difficult to see the stars. In ancient times, however, there were no lights, and stars were a spectacle for everyone to behold on clear nights. People began to see patterns in the positions of the stars and then to identify these patterns with persons, animals, and even monsters. We now have a rich collection of stories whose main characters can be found in patterns of stars in the sky. Some of the star patterns, like the Big Dipper (shown below), are familiar to most of us even if we have not studied astronomy.

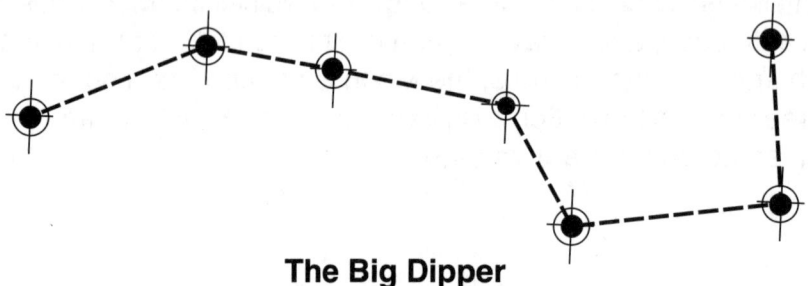

The Big Dipper

Patterns of stars in the sky are correctly called **asterisms**. They are sometimes referred to, however, as **constellations**. The word constellation really stands for a region of the sky. Astronomers have divided the sky into 88 such regions to make it easier to locate sky objects. We will use the word constellation to stand for regions of the sky and for the patterns of stars in those regions.

The constellations and many of the stars contained in them have names. A convenient tool for locating and finding names of asterisms in the sky is the *Edmund Star and Planet Locator*.*

Key Words

Planet: A heavenly body that follows an orbit around the Sun.

Orbit: A path followed by a heavenly body around another one.

Moon: A heavenly body that orbits around the earth or another planet.

Stars: Heavenly bodies that, like the Sun, shine with their own light.

Galaxies: Clusters of large numbers of stars.

Universe: The word we use to mean everything: the earth, the Sun, the planets, the stars, and even the other galaxies.

Asterism: A pattern of stars in the sky.

Constellation: A region of the sky designated by astronomers.

---

* Available from Edmund Scientific Company, Bavrington, New Jersey, 08007

Math in Astronomy   Lesson One: Questions

Name _____ Date _____

Lesson One—Questions

1. A heavenly body that follows an orbit around the Sun is called a

   a) moon   b) universe   c) star   d) planet

2. A region of the sky identified by astronomers and given a name is called a(n)

   a) constellation b) asterism   c) star   d) galaxy

3. A word that is used to mean everything that exists, the earth, the Moon, the Sun, the planets, the stars, and even the galaxies is

   a) moon   b) universe   c) star   d) galaxy

4. A pattern of stars in the sky is called a(n)

   a) constellation b) asterism   c) star   d) galaxy

5. A cluster containing a large number of stars is called a(n)

   a) constellation b) asterism   c) universe   d) galaxy

6. The path followed by a heavenly body around the Sun is called a(n)

   a) moon   b) orbit   c) star   d) planet

7. A heavenly body that orbits around a planet is called a(n)

   a) moon   b) asterism   c) star   d) planet

8. A heavenly body that produces its own light is called a(n)

   a) moon   b) asterism   c) star   d) planet

9. The study of heavenly bodies is called

   a) ecology   b) meteorology   c) orbit science d) astronomy

10. The galaxy in which our Sun and its planets are located is called the

    a) Starry Path   b) Milky Way   c) Giant Spiral   d) Cosmic Shimmer

Lesson One—Math

Numbers in Scientific Notation

In astronomy, we often find ourselves confronted with very big numbers. We need to be able to express the very large distances in the solar system and the even larger distances between the stars in numbers that we can understand. We choose to represent these values in scientific notation.

Scientific notation is based on powers of ten. Before we go any further, let's review some things about powers of ten.

The number ten (10) is ten to the first power (10^1).
The number one hundred (100) is ten to the second power (10^2).
The number one thousand (1,000) is ten to the third power (10^3).
The number ten thousand (10,000) is ten to the forth power (10^4).

There is a pattern here. Numbers that consist of a one (1) followed by several zeros (0) can be written as powers of ten. The power of ten is just the number of zeros in the number. Look for the pattern in the list of numbers below.

$10 = 10^1$
$100 = 10^2$
$1,000 = 10^3$
$10,000 = 10^4$
$100,000 = 10^5$
$1,000,000 = 10^6$
$10,000,000 = 10^7$
$100,000,000 = 10^8$

$1,000,000,000 = 10^9$
$10,000,000,000 = 10^{10}$
$100,000,000,000 = 10^{11}$
$1,000,000,000,000 = 10^{12}$
$10,000,000,000,000 = 10^{13}$
$100,000,000,000,000 = 10^{14}$
$1,000,000,000,000,000 = 10^{15}$

A power of ten can also be thought of as the number of times ten must be multiplied by itself to get the original number. Look for the pattern in the numbers below.

$10 =$ $10 = 10^1$
$100 =$ $10 \times 10 = 10^2$
$1,000 =$ $10 \times 10 \times 10 = 10^3$
$10,000 =$ $10 \times 10 \times 10 \times 10 = 10^4$
$100,000 =$ $10 \times 10 \times 10 \times 10 \times 10 = 10^5$
$1,000,000 =$ $10 \times 10 \times 10 \times 10 \times 10 \times 10 = 10^6$
$10,000,000 =$ $10 \times 10 \times 10 \times 10 \times 10 \times 10 \times 10 = 10^7$
$100,000,000 = 10 \times 10 \times 10 \times 10 \times 10 \times 10 \times 10 \times 10 = 10^8$

Name _____ Date _____

Numbers in Scientific Notation (continued)

A large number can be put into scientific notation by writing it as the first digit of a number, followed by a decimal point, followed by all of the other digits in the number, multiplied by a power of ten. The number below is in scientific notation.

$$8.75 \times 10^8$$

To convert a large number to scientific notation, you must do two things:

1. Move the decimal point in the number to the left so that it is between the left-end digit in the number and the digit that is next to the left end. If the number does not have a decimal point to start with, put one on the extreme right-hand end of the number before beginning.

2. Multiply the number by a power of ten. The power should be equal to the number of digits that you had to pass when you moved the decimal point to the left in step 1.

Let's see how we can convert the number 8415.7 to scientific notation. First write the number.

$$8415.7$$

Next, move the decimal point between the 8 and the 4.

$$8.4157$$

Next, notice that you had to move the decimal point three digits to the left. Multiply by ten to the third power.

$$8.4157 \times 10^3$$

Let's try another one. Convert the number 532,000 to scientific notation. First, write the number. Since there is no decimal point to start with, place one on the right-hand end of the number.

$$532,000.0$$

Move the decimal point between the 5 and the 3.

$$5.32000$$

Then, since you had to move the decimal point five digits to the left, multiply by ten to the fifth power.

$$5.32 \times 10^5$$

Notice that it is not necessary to keep the extra 0s on the right-hand end of the number.

Exercises—Scientific Notation

Write the following numbers as powers of ten.

1. 100 _____   2. 10,000 _____   3. 1,000 _____   4. 1,000,000 _____
5. 100,000 _____   6. 10,000,000 _____

Write the following numbers in scientific notation.

7. 500 _____   8. 4,572 _____   9. 5,321 _____
10. 725,000 _____   11. 95,476.5 _____   12. 5,895,000 _____
13. 345.67 _____   14. 25,150,000,000 _____   15. 675,123.25 _____

Distances

In astronomy we often must deal with distances. We need to write the distance from the earth to the Sun, the distances to the stars, and even the sizes of planets and Moons. The size of a body can sometimes be thought of as distance from one side of it to the other side.

We think of **distance** as how far it is from one place to another. We will call our places points. A **point** can be thought of as a particular spot in space, and can be drawn as a dot on paper. A **line segment** can be drawn between two points.

A •————————————————————————————• B

Line segment between point A and point B.

The distance between the two points is the length of the line segment that extends between them.

Distance must be measured in **units**. If we just say that the distance between two points is 25, that doesn't mean very much. Do we mean 25 inches or 25 miles? The inches or miles would be the **units** of our distance.

We are used to measuring lengths in inches, feet, and miles. In science, however, we usually measure them in centimeters, meters, and kilometers. We use the following abbreviations for these units.

cm = centimeter   m = meter   km = kilometer

These units are related to the more familiar ones that we use as follows:

**There are 2.54 centimeters in one inch.
One meter is 3.28 feet long.
One kilometer is 0.621 miles long.**

One reason that scientists use cm, m, and km is that it is easy to convert from one unit to another.

1 meter = 100 centimeters   1 kilometer = 1,000 meters

To convert from centimeters to meters, multiply by 0.01 m/cm.

265 cm x 0.01 m/cm = 2.65 m

To convert from meters to centimeters, multiply by 100 cm/m.

25 m x 100 cm/m = 2,500 cm

Name _____ Date _____

Distances (continued)

To convert from meters to kilometers, multiply by 0.001 km/m.

5,750 m x 0.001 km/m = 5.75 km

To convert from kilometers to meters, multiply by 1,000 m/km.

325 km x 1,000 m/km = 325,000 m
= 3.25×10^5 m

It is also possible to convert from inches, feet, and miles to centimeters, meters, and kilometers and from centimeters, meters, and kilometers to inches, feet, and miles. We will only worry about one of these conversions, from miles (abbreviated mi) to kilometers and from kilometers to miles.

To convert from miles to kilometers, multiply by 1.61 km/mi.

50 mi x 1.61 km/mi = 80.5 km

To convert from kilometers to miles, multiply by 0.621 mi/km.

150 km x 0.621 mi/km = 93.2 mi

Exercises—Distances

16. 3,000 cm is how many meters?

17. 25 m is how many centimeters?

18. 10,250 cm is how many meters?

19. 15.5 m is how many centimeters?

20. 2.75×10^5 cm is how many meters?

21. 1.5×10^3 m is how many centimeters?

22. 550 km is how many meters?

23. 25,450 m is how many kilometers?

24. 25,000 km is how many meters?

25. 2.52×10^4 m is how many km?

26. 5 km is how many m?

27. 45 m is how many km?

28. 312 km is how many miles?

29. 25 mi is how many kilometers?

30. 5,500 km is how many mi?

31. 2,570 mi is how many km?

32. 1.5×10^3 km is how many mi?

33. 3.25×10^4 mi is how many km?

Name _____ Date _____

Angles

In astronomy, we sometimes need to talk about directions, and to describe directions, we need angles. We will describe angles very simply here by looking first at the intersection of rays. A **ray** can be thought of as a line segment that has one endpoint and extends forever at the other end. The picture below shows a ray.

The intersection of two rays forms an **angle.** The two rays have a common endpoint.

There are several types of angles, and angles can be measured in many ways. We will only be interested in certain types of angles here.

A "square corner" angle is called a **right angle** and a "straight line" angle is called a **straight angle.** A right angle and a straight angle are shown below. An angle smaller than a right angle is an **acute angle,** and an angle larger than a right angle but smaller than a straight angle is called an **obtuse angle.**

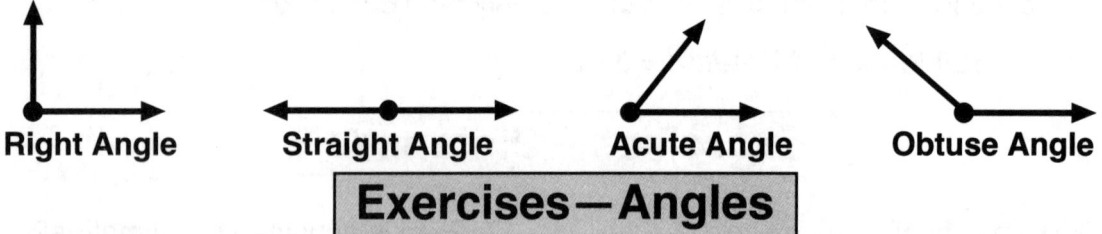

Exercises—Angles

For our purposes, we will measure angles in degrees. There are 180 degrees in a straight angle and 90 degrees in a right angle. (Degrees are often written as °. Ninety degrees would be written as 90°.) Each degree, therefore, is $\frac{1}{180}$ of a straight angle and $\frac{1}{90}$ of a right angle. Your teacher will show you how to use a device called a **protractor** to measure angles. You will need to use the protractor to do the exercises below.

Measure the following angles with a protractor. Classify each angle as an acute angle, a right angle, an obtuse angle, or a straight angle.

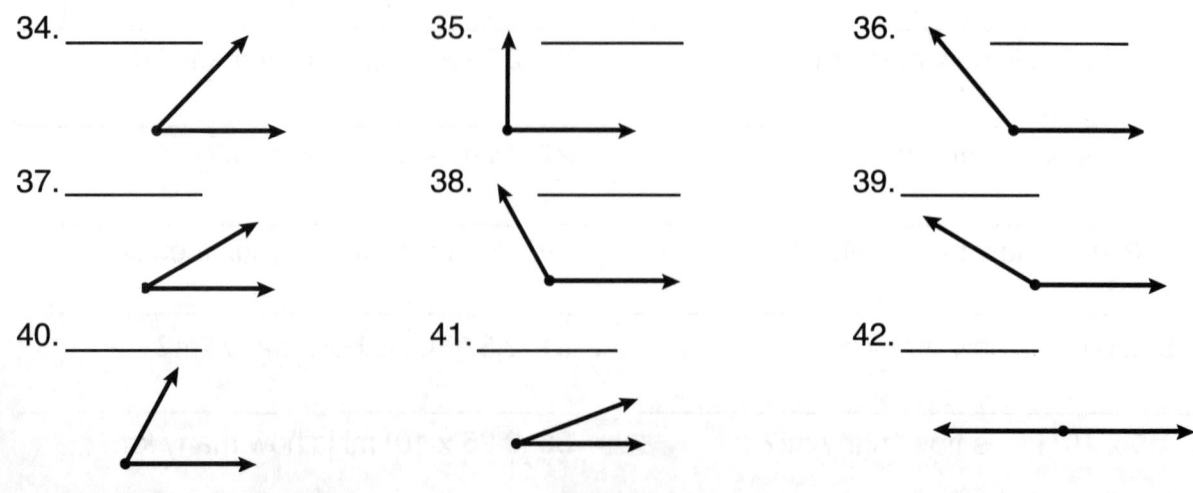

Math in Astronomy

Lesson One—Teacher's Page

Lesson One—Teacher's Page

Comments

1. When multiplying numbers in scientific notation by other smaller numbers, it may be easiest for the students to convert the scientific notation number to ordinary decimal form, multiply it, and then convert the product back to scientific notation.

2. Assume that the kilometers and miles in Mathematical Exercises 28 through 33 are exact values. Multiplying them times the conversion factor would result in a product with three significant digits.

Class Activities

I. Ask students to look up and bring to class some very large numbers (examples: number of nerve cells in the brain (~10^{11}), current world population, and so on) and express them in scientific notation.

II. Ask students to find examples of units that are used to measure distance or length. Examples could come from road signs, products, dispensing units in stores (yards, inches, and so forth), or any other source.

Answers to Questions (page 5)

1. d   2. a   3. b   4. b   5. d   6. b   7. a   8. c   9. d   10. b

Answers to Mathematical Exercises (pages 6–10)

1. 10^2 2. 10^4 3. 10^3 4. 10^6 5. 10^5 6. 10^7

7. 5.0×10^2 8. 4.572×10^3 9. 5.321×10^3
10. 7.25×10^5 11. 9.54765×10^4 12. 5.895×10^6
13. 3.4567×10^2 14. 2.515×10^{10} 15. 6.7512325×10^5

16. 30 m   17. 2,500 cm   18. 102.5 m
19. 1,550 cm   20. 2.75×10^3 m   21. 1.5×10^5 cm
22. 5.5×10^5 m   23. 25.45 km   24. 2.5×10^7 m
25. 25.2 km   26. 5,000 m   27. 0.045 km
28. 193.752 mi   29. 40.25 km   30. 3415.5 mi
31. 4,137.7 km   32. 931.5 mi   33. 5.2325×10^4 km

34. 45°, acute   35. 90°, right   36. 130°, obtuse
37. 30°, acute   38. 120°, obtuse   39. 150°, obtuse
40. 60°, acute   41. 20°, acute   42. 180°, straight

© Mark Twain Media, Inc., Publishers

Lesson Two
The Solar System

Introduction

When someone asks us where our home is, we often respond by telling them the name of the town or the state in which we live. If they ask us what our neighborhood is like, however, we usually describe the houses or people that are closest to our home.

We might respond in a similar way if someone asked us to explain where our home is in the universe. We could say that we live in the Milky Way galaxy on a planet called Earth that is close to a star called the Sun. If we wanted to describe our neighborhood, however, we would have to describe the other bodies in the solar system.

What is the Solar System?

Basically, the **solar system** consists of the Sun and the bodies that are in orbit around it. There are nine large bodies, called planets, that travel in orbits about the Sun. There are also many smaller bodies that orbit the Sun. Some of the smaller bodies would seem very large to us, but they are small when compared to the planets. Other ones are just fragments of rock, chunks of ice, or even grains of dust. The solar system also includes the moons that orbit the planets.

The Planets

The nine planets in the solar system are Mercury, Venus, Earth, Mars, Jupiter, Saturn, Uranus, Neptune, and Pluto. Unlike the Sun, which creates its own light, the planets shine with reflected sunlight. The planets and their orbits are shown in the picture below. The picture is not drawn to scale.

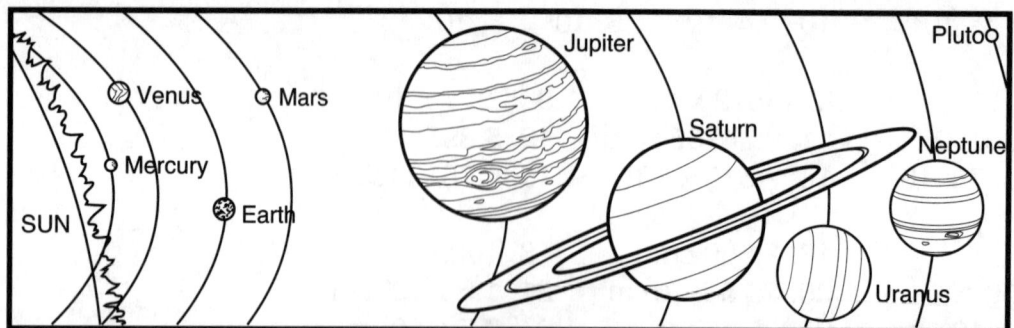

The Planets

As the picture shows, the orbits of the planets are of different sizes so that they do not run into each other. We usually think of the orbits as being circles. Actually, as we will see in the math part of this lesson, the orbits are slightly different from circles. The size of a planet's orbit is usually given as the average distance of the planet from the Sun, a number that is approximately the radius of a circle. As you can see from the table at the end of this section, the sizes of the orbits are very large.

If you could look at the solar system from a viewpoint that was high above the earth's north pole, you would see that the planets all orbit about the Sun in a **counterclockwise** direction. (Counterclockwise is opposite the direction that the hands on a clock rotate.) You would also notice that all of the orbits are almost in the same plane.

Only the closest planets, Mercury, Venus, Mars, Jupiter, and Saturn, are visible to us on Earth without the use of telescopes. The others are too far away and appear too dim. Since the planets travel around the Sun in their orbits, they appear in different areas of the sky at different times. The planets that are closest to the Sun change their positions in the sky most quickly. Planets that are farther from the Sun move slower in their orbits and change their positions more slowly against the background of stars.

Each planet takes a certain amount of time to go completely around the Sun in its orbit. This time is called the planet's **sidereal period** or just its **period.** Planets that are close to the Sun have short sidereal periods and planets that are farther from the Sun have longer ones. The table below shows the periods (in Earth years) and average distances from the Sun (in km) for the planets.

PLANET	AVERAGE DISTANCE FROM SUN (km)	PERIOD (yrs)
Mercury	5.79×10^7	0.24
Venus	1.08×10^8	0.62
Earth	1.50×10^8	1.00
Mars	2.28×10^8	1.88
Jupiter	7.78×10^8	11.9
Saturn	1.43×10^9	29.5
Uranus	2.87×10^9	84.0
Neptune	4.50×10^9	165
Pluto	5.92×10^9	249

Inferior Planets

Planets whose orbits are closer to the Sun than Earth's orbit are called **inferior planets**. Mercury and Venus are inferior planets. Some special orbital positions of these planets are indicated in the picture below.

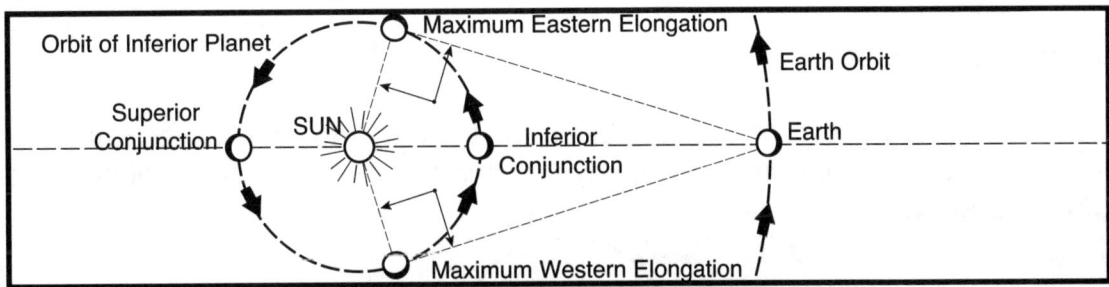

Orbit Positions: Inferior Planets

Inferior conjunction is the position of an inferior planet when it is closest to Earth. It lies along an imaginary line that extends between Earth and the Sun, and the planet is between the earth and the Sun. In this position, the Sun shines on the opposite side of the planet so we do not see it as a bright spot in the sky. Sometimes, however, we can see the planet as a tiny black dot against the bright Sun.

Superior conjunction is the position of an inferior planet when it is farthest from Earth. It is on the opposite side of the Sun from the earth. Although the side facing us is completely illuminated, we cannot see it because the Sun is between us and the planet.

Maximum elongation is the position of an inferior planet when it is at the largest angle from the Sun as viewed by someone on Earth. When it is at maximum eastern elongation, it is seen in the western sky just after sunset by an observer on Earth, and it is often called an **evening star**. When it is at maximum western elongation, it is seen in the eastern sky before dawn and is sometimes called a **morning star.**

Superior Planets

Planets whose orbits are farther from the Sun than Earth's orbit are called **superior planets**. Mars, Jupiter, Saturn, Uranus, Neptune, and Pluto are superior planets. Conjunction and opposition are special orbital positions of superior planets.

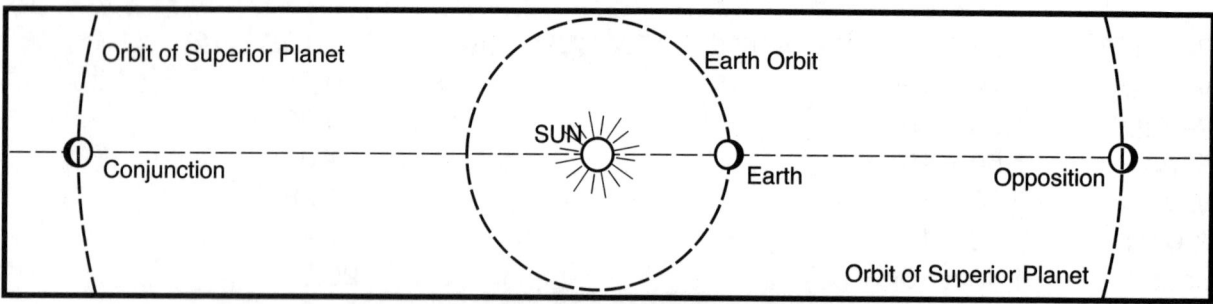

Orbit Positions: Superior Planets

Conjunction is the orbit position of a superior planet when it is farthest from Earth, on the opposite side of the Sun. It is usually impossible to observe the planet at this time.

Opposition is the orbit position of a superior planet when it is closest to Earth. In this position, the planet is on the opposite side of the earth from the Sun. Oppositions are of interest to professional and amateur astronomers because they are the best times to view the planet.

Rotation of the Planets

In addition to traveling in their orbits around the Sun, the planets also rotate. Some of them rotate slowly and others more rapidly. The line about which a planet rotates is called its **axis of rotation.**

A Planet and Its Axis of Rotation

Sizes of the Planets

All of the planets in the solar system are approximately spherical in shape. In other words, they all look like round balls. We usually specify the size of a planet by its diameter. The planets vary greatly in size. The largest planet, Jupiter, has a diameter that is more than eleven times the diameter of the earth. The smallest one, Pluto, has a diameter that is less than two-tenths of Earth's diameter. The data below shows the approximate diameters (in km) of the planets.

PLANET	DIAMETER (km)
Mercury	4,880
Venus	12,100
Earth	12,760
Mars	6,790
Jupiter	142,980
Saturn	120,540
Uranus	51,120
Neptune	49,530
Pluto	2,250

Relative Sizes of Jupiter and Earth

Moons

A **moon** is a heavenly body that orbits around a planet. Our planet, Earth, has one moon. As we will see, however, other planets have several moons, and some have none at all. The planets and the number of moons that they have are listed below.

PLANET	NUMBER OF MOONS (km)
Mercury	none
Venus	none
Earth	1
Mars	2
Jupiter	16
Saturn	20
Uranus	15
Neptune	8
Pluto	1

A Planet and Its Orbiting Moon

Other Bodies in the Solar System

There are bodies other than planets in the solar system. The most interesting ones are asteroids and comets. **Asteroids** are iron or rock bodies that are much smaller than the planets. Most of them travel in orbits between the orbits of Mars and Jupiter, but some of them have orbits that bring them close to Earth. Some astronomers worry that Earth might collide with one of these bodies sometime.

Comets have their origin in the outer portion of the solar system, far outside of the orbit of Pluto. Occasionally they wander into the inner part of the solar system where they can be seen from Earth. Unlike asteroids, comets are composed mainly of dust, gas, and icy materials.

Key Terms

Solar system: The Sun and the bodies that are in orbit around it.

Counterclockwise: A rotation or revolution that is opposite the direction that the hands of a clock rotate.

Sidereal period (period): The time required for an orbiting body to go around the Sun once in its orbit.

Inferior planet: A planet whose average distance to the Sun is less than the average distance from Earth to the Sun.

Inferior conjunction: A position of an inferior planet where the planet is directly between Earth and the Sun. The planet is closest to Earth at that time.

Superior conjunction: A position of an inferior planet where the planet is on the opposite side of the Sun from Earth.

Maximum elongation: A position of an inferior planet where the planet is at its greatest angular distance from the Sun from the point of view of an Earth observer.

Evening star: An inferior planet that is seen in the sky just after sunset.

Morning star: An inferior planet that is seen in the sky just before dawn.

Superior planet: A planet whose average distance to the Sun is greater than the average distance from Earth to the Sun.

Conjunction: A position of a superior planet where the planet is on the opposite side of the Sun from Earth.

Opposition: A position of a superior planet where the planet is on the opposite side of the earth from the Sun. The planet is closest to Earth at that time.

Asteriod: A small, rocky or iron body that is in orbit around the Sun. Most asteroid orbits are between the orbits of Mars and Jupiter.

Comet: A body, composed mainly of ice and dust, that is in an orbit around the Sun.

Math in Astronomy   Lesson Two—Questions

Name _____ Date _____

Lesson Two—Questions

1. The Sun and the bodies that are in orbit around it comprise the

   a) inferior planets   b) solar system   c) galaxies   d) superior planets

2. When viewed from a direction above Earth's North Pole, the planets appear to be circling the Sun in a _____ direction.

   a) clockwise   b) counterclockwise

3. The time that a planet takes to make one complete orbit around the Sun is called the planet's

   a) day   b) hour   c) semi-major axis   d) period

4. A planet whose orbit lies inside Earth's orbit is called a(n)

   a) superior planet   b) inferior planet   c) moon   d) comet

5. A planet whose orbit lies outside Earth's orbit is called a(n)

   a) superior planet   b) inferior planet   c) moon   d) comet

6. At inferior conjunction, an inferior planet would be

   a) between Earth and the Sun
   b) on the other side of the Sun from the earth
   c) at its greatest angular distance from the Sun
   d) on the opposite side of the Sun from Earth

7. At maximum elongation, an inferior planet would be

   a) between Earth and the Sun
   b) on the other side of the Sun from the earth
   c) at its greatest angular distance from the Sun
   d) on the opposite side of the Sun from Earth

8. At superior conjunction, an inferior planet would be

   a) between Earth and the Sun
   b) on the other side of the Sun from the earth
   c) at its greatest angular distance from the Sun
   d) on the opposite side of the Sun from Earth

Name _____ Date _____

Lesson Two—Questions (continued)

9. At conjunction, a superior planet would be

   a) between Earth and the Sun
   b) on the other side of the Sun from the Earth
   c) at its greatest angular distance from the Sun
   d) on the opposite side of the Sun from Earth

10. At opposition, a superior planet would be

    a) between Earth and the Sun
    b) on the other side of the Sun from the earth
    c) at its greatest angular distance from the Sun
    d) on the opposite side of Earth from the Sun

11. An inferior planet that is seen in the morning sky is called a(n)

    a) morning star   b) evening star   c) dark star   d) moon

12. An inferior planet that is seen in the evening sky is called a(n)

    a) morning star   b) evening star   c) dark star   d) moon

13. A small, rocky or iron body in orbit around the Sun, usually between the orbits of Mars and Jupiter is called a(n)

    a) comet   b) moon   c) asteroid   d) inferior planet

14. A body, composed mainly of ices and dust, that is in orbit around the Sun is called a(n)

    a) comet   b) moon   c) asteroid   d) inferior planet

15. The two inferior planets are

    a) Mars and Venus   b) Mercury and Venus
    c) Mercury and Jupiter   d) Mars and Saturn

Lesson Two—Math

Time

Time is something that most of us take for granted. We can always find the time of day by looking at a clock or wristwatch. We can keep track of the day of the month and the month of the year by looking at a calendar.

Astronomy provides us with the means to tell time. Both the length of a day and the season of a year are based on astronomic events. We will look at how a day is defined in Lesson Four, but we can investigate a year right now.

One kind, called a **sidereal year,** was described in this lesson. It is just the amount of time that it takes Earth to make one complete orbit around the Sun. This period can be expressed in days. It is approximately 365.26 days.

The first day of spring occurs when the earth is at a certain orientation in its orbit. Because of a slight wobble in the axis of rotation of the earth, the time between the beginnings of two first days of spring is slightly shorter than a sidereal year. It is approximately 365.24 days. We call it a **tropical year.**

Since there is not an even number of days in either a sidereal or a tropical year, we use a calendar that tries to keep the number of days in a year almost the same each year. The calendar also tries to keep the first day of spring falling on about the same date each year. Here is how this calendar works (for recent years).

We have two kinds of years, **regular years** and **leap years.** Regular years have 365 days, and leap years have 366 days.

Years (except century years) that are evenly divisible by four (4) (no fraction or remainder when you divide them by four) are leap years. Century years (1900, 2000, and so forth) are leap years if they are evenly divisible by 400. All other years are regular years.

Let's see if 1950 was a leap year or a regular year. $\frac{1950}{4} = 487.5$

Since 4 does not divide evenly into 1950, it is a regular year.

Let's check to see if 1964 was a leap year or a regular year. $\frac{1964}{4} = 491$

Since 4 divides evenly into 1964, it is a leap year.

Will 2100 be a leap year or a regular year? This is a century year, so we will divide it by 400.

$$\frac{2100}{400} = 5.25$$

Since it is not evenly divisible by 400, 2100 is a regular year.

Math in Astronomy   Lesson Two—Math: Time/The Astronomical Unit

Name _____ Date _____

Exercises—Time: Calendar

1. Was the year 1976 a leap year? _____

2. Was the year 1958 a leap year? _____

3. Was the year 1932 a leap year? _____

4. Was the year 1900 a leap year? _____

5. Will the year 2000 be a leap year? _____

The Astronomical Unit

The distances between bodies in the solar system are very large when they are expressed in kilometers. It would be better if we could use a larger unit to write these distances so that they would be easier for us to understand. Astronomers have created such a unit, the **Astronomical Unit,** abbreviated AU.

An Astronomical Unit is equal to the average distance between Earth and the Sun (approximately 1.5×10^8 km.) The average distances of the planets from the Sun, in Astronomical Units (AU's) are given in the following table. Notice how the distances have more meaning than the ones given earlier in this lesson, because they use smaller numbers that are more easily comprehended at a glance.

PLANET	AVERAGE DISTANCE FROM THE SUN
Mercury	0.39 AU
Venus	0.72 AU
Earth	1.00 AU
Mars	1.52 AU
Jupiter	5.20 AU
Saturn	9.54 AU
Uranus	19.18 AU
Neptune	30.06 AU
Pluto	39.44 AU

Math in Astronomy   Lesson Two—Math: Astronomical Units

Name _____ Date _____

Exercises—Astronomical Units

The following problems will give you some practice doing simple arithmetic with AU's. These calculations would be much more difficult if the distances were expressed in kilometers.

6. How far (in AU's) is Earth from Mars when Mars is in opposition? See the diagram below that shows the positions of Earth and Mars. _____

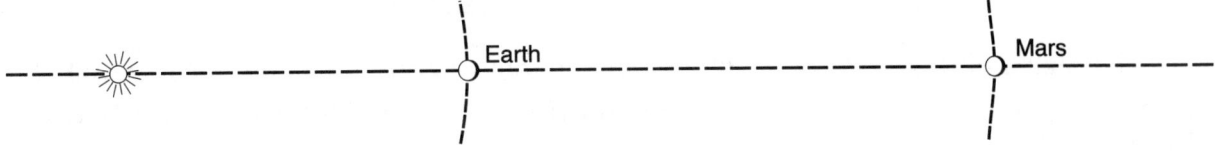

7. How far (in AU's) is Earth from Uranus when Uranus is in opposition? See the diagram below that shows the positions of Earth and Uranus. _____

8. How far (in AU's) is Earth from Venus when Venus is in inferior conjunction? See the diagram below that shows the relative positions of Earth and Venus. _____

9. How far (in AU's) is Earth from Mercury when Mercury is in superior conjunction? See the diagram below that shows the relative positions of Earth and Mercury. _____

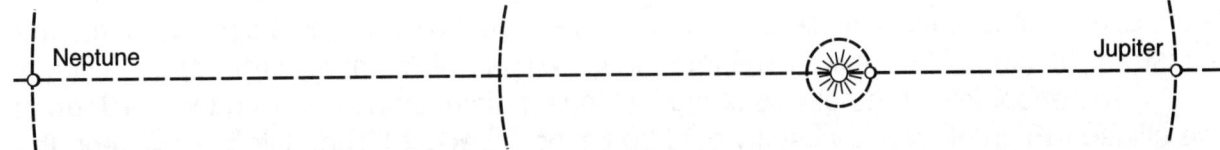

10. How far (in AU's) is Neptune from Jupiter when the two planets are in the positions shown in the diagram below. _____

© Mark Twain Media, Inc., Publishers   21

Kepler's Laws

In the 1600s an astronomer, **Johannes Kepler,** announced three laws that described how planets moved in their orbits around the Sun. These three laws give us a better understanding of the motion of planets and how mathematics can help us predict the motion of the planets. Kepler gave us three laws, but we will only be interested in two of them, the First Law and the Third Law.

Kepler's First Law: Ellipses

Kepler's First Law states that the planets move in orbits that are ellipses with the Sun at one focus of the ellipse. To understand this law, we need to know something about ellipses.

An **ellipse** is a figure that looks like a stretched out circle. The easiest way to understand an ellipse is to draw one. The picture below shows how to draw an ellipse with a pencil, a length of string, and two thumbtacks.

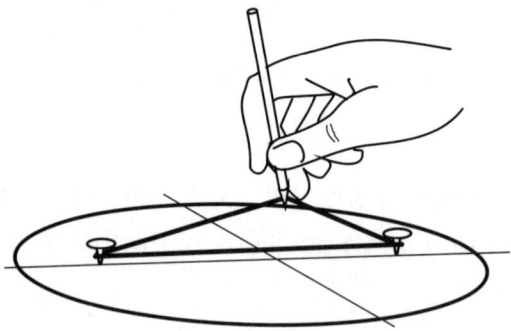

Place a piece of paper on a piece of cardboard and draw a horizontal line across the paper. Then insert the thumbtacks at two spots along the line. Tie the two ends of the string together and place the created loop of string over the thumbtacks. Then trace the pencil completely around the thumbtacks, always keeping the string tight. You will draw a figure like the one shown below.

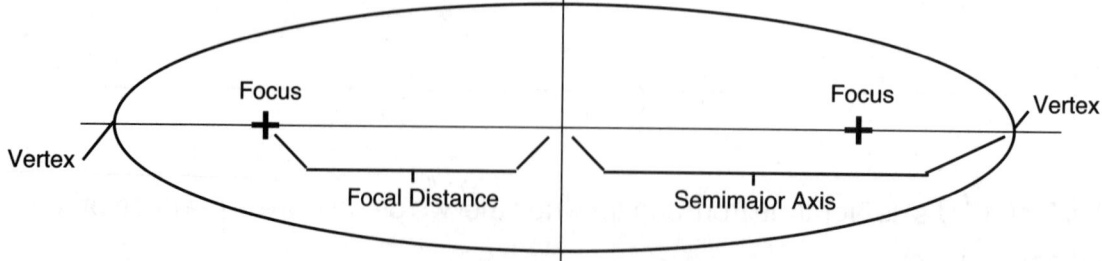

The two thumbtack holes are called the **focus points** of the ellipse. Each focus point is the same distance from the center of the ellipse. If the two focus points are far apart, the ellipse will be long. If they are close together, the ellipse will be almost round, like a circle.

The endpoints of the ellipse along its longest dimension are called the **vertices** of the ellipse. (Each of them by itself would be called a **vertex**.) The distance between the center of the ellipse and one of the vertices is called the **semimajor axis** of the ellipse. The

Math in Astronomy   Lesson Two—Math: Kepler's First Law: Ellipses

Name _____   Date _____

distance from the center of the ellipse to one of the focus points is called the **focal distance**.
An ellipse property called the **eccentricity** measures how long and skinny or how round an ellipse is.

$$\text{eccentricity} = \frac{\text{focal distance}}{\text{semimajor axis}}$$

The eccentricity can be a number between 0 and 1. If it is close to 1, the ellipse will be long and skinny. If it is close to 0, the ellipse will be almost a circle.

Exercises— Ellipses

Calculate the eccentricities of the following ellipses by measuring the semimajor axis and the focal distance and then dividing the focal distance by the semimajor axis.

11. _____

12. _____

13. _____

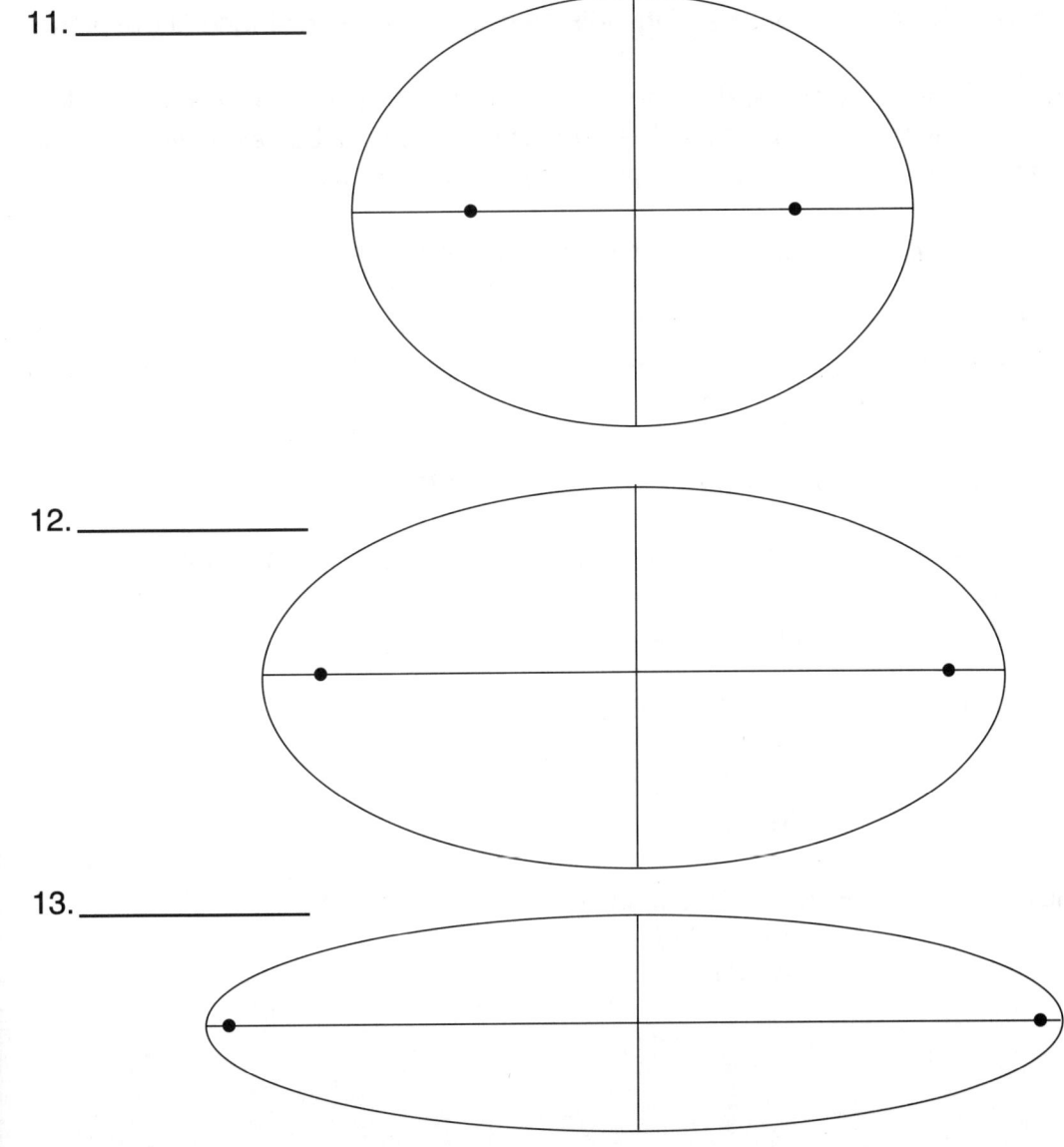

© Mark Twain Media, Inc., Publishers   23

Kepler's First Law: Ellipses (continued)

The picture below shows a planet in an orbit around the Sun. The orbit is an ellipse with the Sun at one of the focus points.

The eccentricity of the ellipse in this picture has been exaggerated. The orbits of the planets have very small eccentricities and are almost circular.

The length of the semimajor axis of the ellipse is the average distance of the planet from the Sun.

The spot in a planet's elliptical orbit where it comes closest to the Sun is called **perihelion.** The distance from the planet to the Sun at perihelion can be calculated from the eccentricity of the planet's orbit and the length of the semimajor axis.

perihelion distance = semimajor axis x (1 - eccentricity)

The spot in a planet's elliptical orbit where it is farthest from the Sun is called **aphelion.** The distance from the planet to the Sun at aphelion can also be calculated from the eccentricity of the planet's orbit and the length of the semimajor axis.

aphelion distance = semimajor axis x (1 + eccentricity)

Let's calculate the perihelion distance and the aphelion distance for the planet Jupiter. Jupiter's average distance from the Sun (equal to its semimajor axis length) is 5.20 AU, and the eccentricity of its orbit is 0.048.

perihelion distance = 5.20 x (1 - 0.048)
= 5.20 x (1.000 - 0.048)
= 5.20 x 0.952
= 4.95 AU

aphelion distance = 5.20 x (1 + 0.048)
= 5.20 x 1.048
= 5.45 AU

Math in Astronomy   Lesson Two—Math: Kepler's First Law: Ellipses

Name _____ Date _____

The eccentricities of the planets' orbits and the average distances of the planets from the Sun are shown below. Use this information to answer the following questions.

Planet	Eccentricity	Planet	Average Distance From Sun
Mercury	0.206	Mercury	0.39 AU
Venus	0.007	Venus	0.72 AU
Earth	0.017	Earth	1.00 AU
Mars	0.093	Mars	1.52 AU
Jupiter	0.048	Jupiter	5.20 AU
Saturn	0.056	Saturn	9.54 AU
Uranus	0.046	Uranus	19.18 AU
Neptune	0.010	Neptune	30.06 AU
Pluto	0.248	Pluto	39.44 AU

Exercises—Kepler's First Law

14. Calculate the perihelion and aphelion distances for the planet Venus.

_____

_____

15. Calculate the perihelion and aphelion distances for the planet Earth.

_____

_____

16. Calculate the perihelion and aphelion distances for the planet Pluto. Is the perihelion distance less than the average distance of Neptune to the Sun?

_____

_____

Math in Astronomy   Lesson Two—Math: Kepler's Second Law/Kepler's Third Law

Name _____ Date _____

Kepler's Second Law

Kepler's Second Law is expressed in a complicated way. As far as we are concerned, it means that planets move faster in their orbits when they are closer to the Sun than when they are farther away from the Sun. We will not do any mathematics with this law.

Kepler's Third Law

Kepler's Third Law states that the square of the period (in years) of a body orbiting the Sun in the solar system is equal to the cube of its average distance from the Sun (semimajor axis).

$$(\text{period})^2 = (\text{semimajor axis})^3$$

To **square** a number, you just multiply it times itself. For example, 3 squared (3^2) is $3 \times 3 = 9$. To **cube** a number, you multiply it times itself one more time. For example, 3 cubed (3^3) is $3 \times 3 \times 3 = 27$.

Let's check to see if Kepler's Third Law is true for the planet Neptune. The square of the period (in years) of Neptune is $164.8^2 = 27{,}160$. The cube of the average distance to the Sun is $30.06^3 = 27{,}160$. It looks like Kepler's Third Law is valid. (Notice that both of these numbers have been rounded off to the same number of digits that the original values had.)

Exercises—Kepler's Third Law

Use the sidereal period information from page 13 and the average distance from the sun information on page 20 to answer the following questions.

17. See if Kepler's Third Law applies to the planet Pluto. Show your work. _____

18. See if Kepler's Third Law applies to the planet Saturn. Show your work. _____

Kepler's Third Law also holds for bodies other than planets. Suppose some body is orbiting the Sun with a period of 8 years. How can we calculate its average distance from the Sun?

The square of its period is $8^2 = 64$. What number cubed is 64? Trial and error shows that 4 cubed is 64. Therefore, the average distance of this body from the Sun is 4 AU.

NOTE: When we find the number that, when cubed, is equal to another number, we are finding the **cube root** of that other number. In the above example, we found the cube root of 64 to be 4. It was easy because the answer turned out be a whole number. Other problems may not be so easy. Your teacher may show you how to find a cube root on your calculator.

Name _____ Date _____

Exercises—Kepler's Third Law (continued)

19. An asteroid is orbiting the Sun with a period of 27 years. What is the average distance of the asteroid from the Sun?

_____

20. A comet is orbiting the Sun with a period of 125 years. What is the average distance of the comet from the Sun?

_____

Suppose that we know that the average distance of a body from the Sun is 9 AU. How can we find its period?

The cube of the distance is $9^3 = 729$. What number squared is 729? Trial and error shows that 27 squared is 729.

Therefore, the period of the body is 27 years.

NOTE: When we find the number that, when squared, is equal to another number, we are finding the **square root** of that other number. In the example above, we found the square root of 729. It was not difficult because the answer was a whole number. For more complicated square roots, your teacher may ask you to use the square root key on a calculator.

Exercises—Kepler's Third Law

21. An asteroid is traveling around the Sun in an orbit whose semimajor axis is 4 AU. What is the period of the asteroid's orbit?

_____

22. A comet is traveling around the Sun in an orbit whose semimajor axis is 50 AU. What is the period of the comet's orbit?

_____

Scaling

Sometimes it is useful to make scale models of things. You may have built a model airplane that was a scale model of the real thing or had a doll that was a scale model of a person. The secret of a good scale model is finding the **scale.**

A scale is found by dividing some length or distance that you know or want on the model by the actual distance on the real thing.

$$\text{scale} = \frac{\text{length on model}}{\text{actual length}}$$

Once a scale is known, you can find all other lengths on the model by multiplying the actual length on the real thing by the scale.

length on model = scale x actual length

To see how this works, let's make a scale model of the solar system. We will choose to make a model that places a model of Mars 10 centimeters from a model of the Sun. Since we know that Mars is really 1.52 AU from the Sun, our scale will be

$$\text{scale} = \frac{10 \text{ cm}}{1.52 \text{ AU}} = 6.6 \text{ cm/AU}$$

Now, let's calculate the distances of all of the other model planets to our model Sun.
Mercury is actually 0.39 AU from the Sun. On the model, it would be
  distance = 6.6 cm/AU x 0.39 AU = 2.6 cm from the model Sun.
Venus is actually 0.72 AU from the Sun. On the model, it would be
  distance = 6.6 cm/AU x 0.72 = 4.8 cm from the model Sun.
Earth is 1 AU from the Sun. On the model, it would be
  distance = 6.6 cm/AU x 1 = 6.6 cm from the model Sun.
We already know that **Mars** is 10 cm from the model Sun.

Exercises—Scaling

23. On your own paper, complete the model by calculating how far the models of **Jupiter, Saturn, Uranus, Neptune,** and **Pluto** would be from the model Sun. Use the average distance chart on page 20.

24. Make a drawing of the model solar system that you calculated in the above exercise. Find a long (at least 3 m) sheet of paper (brown wrapping paper would do nicely), and draw a big circle for the Sun and little circles for the planets. Use a meter stick to measure off the right distance of each planet from the Sun. Measure from the center of the Sun to the center of the spot where the planet is to be drawn.

25. Make another model of the solar system. This time choose a model that places the model of **Pluto** 100 cm from the model Sun.

Lesson Two—Teacher's Page

Comments

Your students can easily find cube roots on calculators even if the calculators do not have a built-in cube root finder. To find a cube root on a standard calculator: (1) type in the number that you want a cube root of, (2) press the Y^x key, (3) type in 0.3333333333 (or a decimal point and as many threes as your calculator will accept), and (4) press the = key.

Class Activity

After the students have completed the scaling exercises, ask someone to bring in a basketball. If the basketball represents the earth, find the diameters of all of the other planets in "basketballs."

Answers to Questions (pages 17–18)

1. b 2. b 3. d 4. b 5. a 6. a 7. c 8. b 9. b 10. d 11. a 12. b 13. c 14. a 15. b

Answers to Mathematical Exercises (pages 19–29)

1. Yes 2. No 3. Yes 4. No 5. Yes

6. 0.52 AU 7. 18.18 AU 8. 0.28 AU 9. 1.39 AU 10. 35.26 AU

11. 0.59 12. 0.86 13. 0.97

14. perihelion = 0.71 AU aphelion = 0.73 AU
15. perihelion = 0.98 AU aphelion = 1.02 AU
16. perihelion = 29.7 AU aphelion = 49.2 AU Yes

17. Yes 18. Yes

19. 9 AU 20. 25 AU

21. 8 years 22. 354 years

23. Jupiter: 34 cm; Saturn: 63 cm; Uranus: 127 cm; Neptune: 198 cm; Pluto: 260 cm

25. Scale = 2.5 cm/AU
    Mercury: 0.98 cm; Venus: 1.8 cm; Earth: 2.5 cm; Mars: 3.8 cm; Jupiter: 13 cm; Saturn: 24 cm; Uranus: 48 cm; Neptune: 75 cm

Lesson Three
Mercury and Venus: The Inferior Planets

Introduction

The two planets that are closest to the Sun are Mercury and Venus. As we saw in Lesson Two, they are called inferior planets because their orbits are closer to the Sun than the orbit of Earth. Of the two planets, Mercury is the harder one to observe. It is small, and it is never found very far away from the Sun. Therefore, it is only seen in the western sky soon after sunset or in the eastern sky soon before sunrise.

Venus is easier to see than Mercury. Like Mercury, however, it is closer to the Sun than the earth and can only be observed in the evening sky after sunset or in the morning sky before sunrise. It is a little farther from the Sun than Mercury so the length of time that it is visible after sunset or before sunrise is longer than for Mercury. Unlike Mercury, Venus can be a very bright object in the sky—only the Sun and the Moon outshine it. This brightness sometimes earns it the name "morning star" or "evening star." At its brightest, Venus can even cast a shadow.

We will see how to find the longest times that Mercury and Venus are visible in the last section of this lesson.

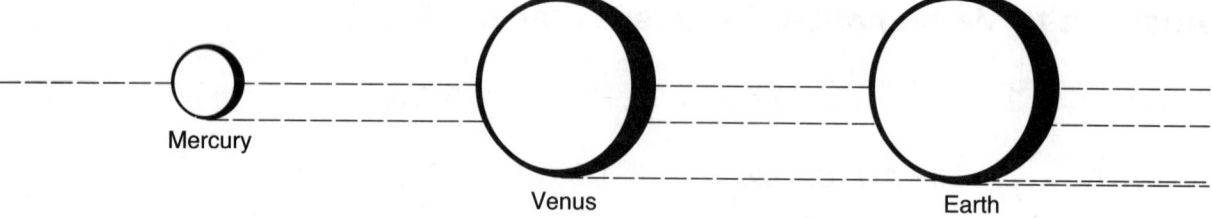

Relative Sizes of Mercury, Venus, and Earth

Mercury

Mercury is only 0.39 AU from the Sun, closer than any of the other planets. It zooms around the Sun in its orbit in only 88 days (0.24 years). A "Mercury year" is only about 88 days long.

NOTE: Whenever we use the word **year** we mean an Earth year. If we refer to a year with respect to another planet, we will use that planet's name in front of the word year (for instance: Mercury year).

The **rotation period** is the time it takes a planet to rotate once about its axis of rotation. The rotation period that we will most often use is the **sidereal rotation period,** the amount of time required to rotate once with respect to the stars. If the planet is also moving in its orbit around the Sun, it takes a different amount of time to rotate once with respect to the Sun. This time is called the **solar rotation period.** The diagram on the next page shows why the sidereal rotation period and the solar rotation period are different.

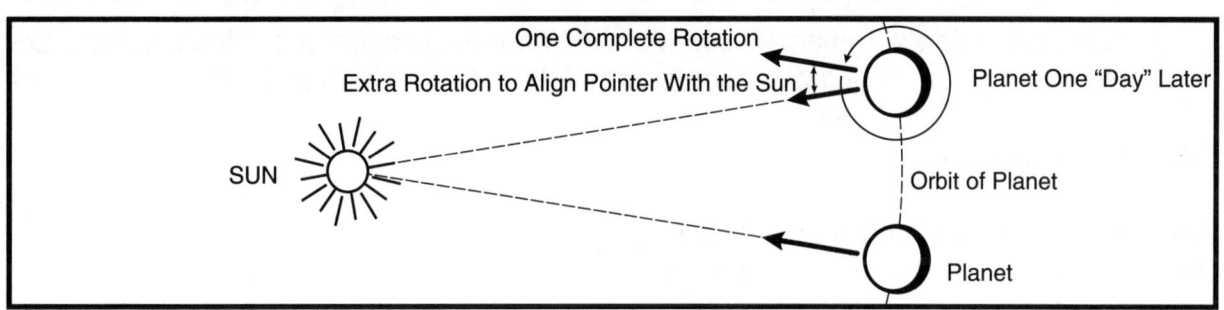

Sidereal and Solar Rotation Periods

Suppose a planet started out with an arrow pointing at the Sun as is shown in the picture. The planet rotates all the way around once, as indicated by the thin line. In the mean time, however, the planet has moved some distance in its orbit so that the arrow does not point to the Sun. The planet must rotate a little more, indicated by the thick line, to make the arrow point to the Sun again. The time taken to rotate around once is the **sidereal rotation period,** and the time taken to rotate around once plus go a little farther to point the arrow at the Sun again is called the **solar rotation period.** If the planet rotates counterclockwise, the solar rotation period will be longer than the sidereal rotation period. If the planet rotates clockwise, the solar period will be shorter.

The sidereal rotation period of Mercury is about 59 days. Because of Mercury's rapid motion in its orbit, a solar rotation period is almost three times this long. This means that it would be a long time between sunrise and sunset (almost 90 days). Think about it! What if you had to go to school for a long day like that?

NOTE: Whenever we use the word **day** we mean an Earth day (24 hours). If we refer to a day with respect to another planet, we will use that planet's name in front of the word day (for instance: Mercury day).

The attraction of gravity (see math section) on the surface of Mercury is only 0.38 what it is on Earth. This means that you would weigh a lot less and that you could jump a lot higher on Mercury than on Earth.

We know something about the surface of Mercury because the United States sent a space probe there in 1974 and 1975. The spacecraft, called Mariner 10, flew by Mercury several times and returned measurements of many types and pictures of the planet's surface. The surface of Mercury is rocky and contains craters, mountains, ridges, valleys, and plains. The craters were caused by the impact of large, heavy bodies that smashed into its surface during its early history. The mountains, ridges, and valleys were also probably caused by these impacts. The plains may be areas where liquid rock, called **lava,** flowed and then became solid a long, long time ago.

Mercury is very close to the Sun, and we might expect it to be very hot. This is true for the sunlit side of the planet, but the side away from the Sun can be very cold. We will investigate temperature and the temperatures of Mercury and Venus in the math portion of this lesson.

Mercury has no atmosphere. Its gravity is so small, and its sunlit surface is so hot, that gases like the air on Earth cannot stay on Mercury. Therefore, it is an airless world that has no wind, no clouds, and no rain.

Math in Astronomy   Lesson Three: Mercury and Venus

The table below summarizes some of the important properties of Mercury. Approximate values are given. (You have seen some of them before in Lesson Two.)

Mercury Properties

Average distance from the Sun	0.39 AU
Orbit period	0.24 years
Orbit eccentricity	0.206
Inclination of equator to orbit	0° (See math section)
Sidereal rotation period	59 days
Diameter	4,880 km
Mass	3.30×10^{23} km (See math section)
	0.056 Earth masses (See Lesson Four)
Average density	5,430 kg/m^3 (See math section)
Gravity	0.38 of Earth (See math section)

Venus

Venus is the second closest planet to the Sun. It takes only 225 days (0.62 years) to complete one orbit around the Sun. A Venusian year is only 225 days long.

The sidereal rotation period of Venus is about 243 days. A rotation with respect to the Sun is a little less than 117 days long. Therefore, a day, from sunrise to sunset, on Venus would be about 58 hours.

Venus' rotation is different from the rotations of the other planets in the solar system. When viewed from a direction above Earth's North Pole, most of the planets rotate counterclockwise, the same way that they revolve in their orbits. Venus, however, rotates clockwise. Its rotation is said to be **retrograde.**

It is very similar to Earth in size and in gravitational attraction. Venus's diameter is about 0.95 the diameter of Earth, only about 690 km smaller than Earth's. The gravitational attraction on Venus is about 0.90 of what it is on Earth. You would weigh only a little less on the surface of Venus than you would on the surface of the Earth.

Thanks to several space probes sent out by the United States and the former Soviet Union, we know quite a bit about Venus. The United States has visited Venus with many unmanned spacecraft that have flown by the planet, orbited it, and even plunged into its atmosphere. The latest of these, the *Magellan* probe, orbited Venus and used radar to make a detailed map of the surface. Earlier, the Soviet Union had sent several spacecraft that actually landed softly on the surface of Venus and sent back measurements and pictures.

The surface of Venus is not visible to us through telescopes because of a thick layer of clouds that covers the planet. The visiting spacecraft, however, have given us a good picture of the planet. The surface is rocky and dry. Almost two thirds of it is hilly plains, about a fourth of it is highlands, and the rest is occupied by volcanic mountains. The highest mountain on Venus, **Maxwell Montes,** is about 12 km high. For comparison, Mount Everest, the highest mountain on Earth, is only 8.8 km high. There are also craters on the surface of Venus. Like the ones on Mercury, they were probably formed long ago by the impacts of large, rocky bodies. The temperature on the surface of Venus is hot enough to melt lead. The planet would not be a nice place for a picnic.

The atmosphere of Venus is thick and heavy and composed mostly of carbon dioxide. It presses down on Venus 100 times as much as our atmosphere presses down on us. We would be crushed by the heavy atmosphere if we could somehow get to the surface of Venus. In addition, we would be unable to breath the carbon dioxide "air."

The clouds that cover up the planet are thought to be composed of sulfuric acid and sulfur particles.

The table below summarizes some of the important Venusian properties. Approximate values are given.

Venus Properties

Average distance from the Sun	0.72 AU
Orbit period	0.62 years
Orbit eccentricity	0.0068
Inclination of the equator to orbit	177° (See math section)
Sidereal rotation period	243 days (retrograde)
Diameter	12,100 km
Mass	4.87×10^{24} km (See math section)
	0.82 Earth masses (See Lesson Four)
Average density	5,250 kg/m³ (See math section)
Gravity	0.90 of Earth (See math section)

Appearance of Venus or Mercury Through Binoculars and Telescopes

Mercury and Venus are both inferior planets. Therefore, they appear in crescent and "quarter Moon" shapes at certain times when we view them. Mercury is farthest away and appears only as a dot in the sky to the naked eye. A good telescope is usually required to show the shape of Mercury, but even high-powered binoculars or low-powered telescopes can reveal the shape of Venus.

At maximum elongation, Mercury or Venus is in the configuration shown in the picture below. We will describe the view as though the planet is Venus.

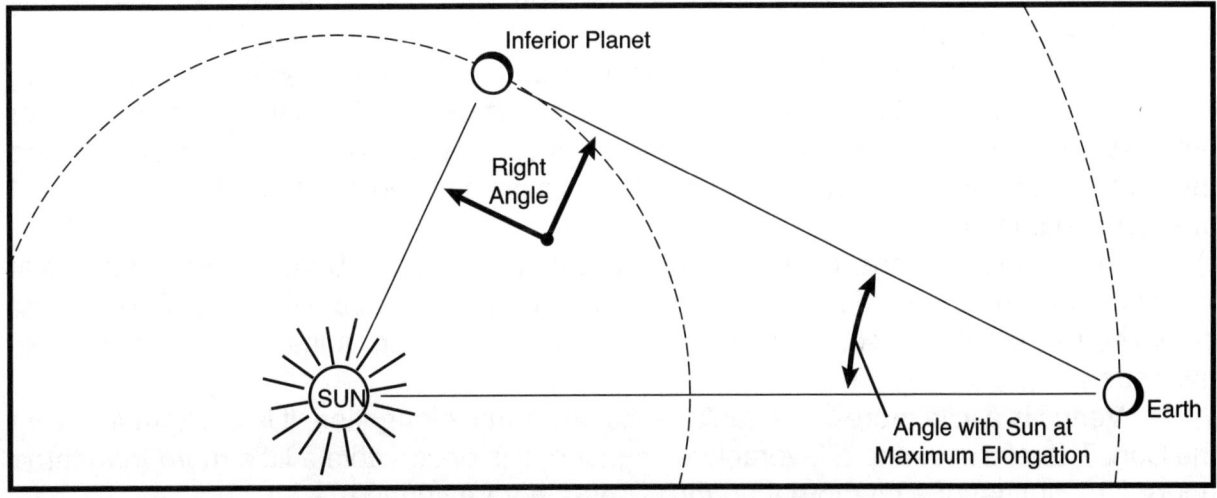

An Inferior Planet at Maximum Elongation

At maximum elongation, Venus would appear like a first or last quarter Moon, in the shape shown below. Venus (or Mercury) would be brightest about halfway between maximum elongation and inferior conjunction. At that position, Venus (or Mercury) would have a crescent shape like the picture below.

Inferior Planet at Maximum Elongation

Inferior Planet at Maximum Brightness

At positions in its orbit between maximum brightness and inferior conjunction, Venus (or Mercury) would appear larger but in a narrower crescent. Finally, at inferior conjunction, the planet would be between the earth and the Sun. The dark backside of the planet would face the earth, and it would not be observable.

Between maximum elongation and superior conjunction, more and more of the disk of the planet would be observable, but it would be smaller. Finally, at superior conjunction, the entire disk of the planet would be observable from Earth. Unfortunately, it would be lost in the glare of the Sun.

Many people feel that the best time to see an inferior planet is at maximum elongation. There are **favorable maximum elongations** and **unfavorable maximum elongations.** A favorable maximum elongation occurs when the position of the earth and the inferior planet is such that it sets in the west or it rises above the Sun in the east, almost at a right angle to the horizon.

At maximum elongation, Mercury is at about 28° from the Sun. The earth rotates at 15° per hour. Therefore, Mercury is never visible more than about two hours after sunset or two hours before sunrise. It is always in the western sky near sunset or the eastern sky near sunrise.

Venus is a little more observable. At its maximum elongation, it is at about 47° from the Sun. Therefore, during a favorable elongation, it is observable a little more than three hours after sunset or a bit more than three hours before sunrise.

Greatest Elongations of Mercury

Evening Sky	Morning Sky
Nov. 28, 1997	Jan. 6, 1998
March 20, 1998*	May 4, 1998
July 17, 1998	Aug. 31, 1998*
Nov. 11, 1998	Dec. 20, 1998
March 3, 1999*	April 16, 1999
June 28, 1999	Aug. 14, 1999
Oct. 24, 1999	Dec. 3, 1999
Feb. 15, 2000	March 28, 2000
June 9, 2000	July 27, 2000
Oct. 6, 2000	Nov. 15, 2000
Jan. 28, 2001	March 11, 2001
May 22, 2001	July 9, 2001

Greatest Elongations of Venus

Evening Sky	Morning Sky
Nov. 6, 1997	March 27, 1998
June 11, 1999	Oct. 30, 1999
Jan. 17, 2001	June 8, 2001

* Most favorable elongations

Key Terms

Sidereal rotation period: Period of rotation of a planet with respect to the stars.

Lava: Liquid rock on the surface of a planet.

Retrograde: Clockwise rotation of a planet as viewed from a vantage point above Earth's North Pole.

Maxwell Montes: Highest mountain on Venus. Twelve kilometers high.

Favorable maximum elongation: Maximum elongation of an inferior planet when the earth and the planet are in a configuration such that the planet rises or sets at almost a right angle to the horizon.

Unfavorable maximum elongation: Maximum elongation of an inferior planet when the earth and the planet are in a configuration such that the planet rises or sets at an acute angle with the horizon.

Math in Astronomy   Lesson Three—Questions

Name _____   Date _____

Lesson Three—Questions

1. The period of rotation of a planet with respect to the stars is called the planet's

   a) synodic rotation period   b) sidereal rotation period
   c) semimajor axis   d) year

2. Liquid rock on the surface of a planet is called

   a) plastic   b) asphalt   c) drip   d) lava

3. When viewed from a vantage point above Earth's North Pole, a planet that is rotating in the retrograde direction would be rotating

   a) clockwise   b) counterclockwise

4. A favorable maximum elongation of an inferior planet occurs when

   a) the planet rises or sets at a right angle with the horizon
   b) the planet rises or sets at an acute angle with the horizon

5. The semimajor axis of Mercury's orbit is 0.39 AU, and the semimajor axis of Venus' orbit is 0.72 AU. How far is Mercury from Venus when they are in the positions shown below?

   a) 1.39 AU   b) 1.72 AU   c) 0.33 AU   d) 1.11 AU

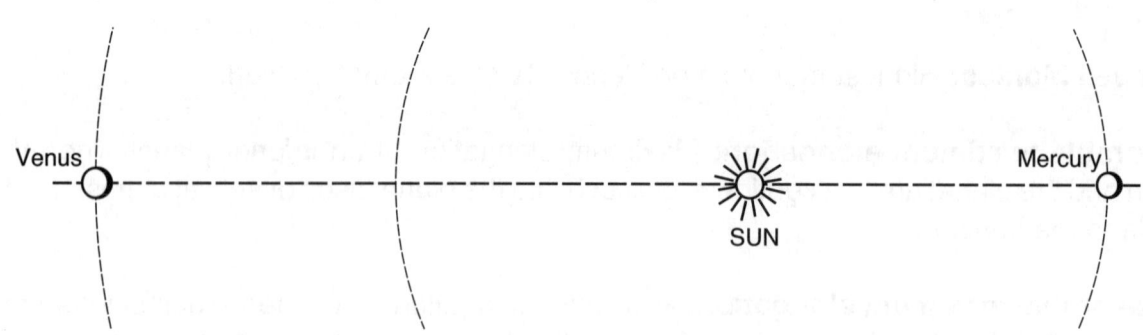

6. The semimajor axis of Mercury's orbit is 0.39 AU and the eccentricity of its orbit is 0.206. What is Mercury's perihelion distance?

   a) 0.47 AU   b) 0.31 AU   c) 0.13 AU   d) 0.29 AU

7. Using the data from question 6, what is Mercury's aphelion distance?

   a) 0.47 AU   b) 0.31 AU   c) 0.13 AU   d) 0.29 AU

Name _____ Date _____

Lesson Three—Math

Mercury Years and Venusian Years

Mercury's year is only 0.24 of an Earth year. What would be your age in Mercury years? You can find out by dividing your age by 0.24 years per Mercury year.

For example, suppose someone was 36 years old on Earth. His or her age in Mercury years would be

Mercury age = $\frac{36}{0.24}$ = 150 Mercury years

Pretty old, huh? Remember that, biologically, you would still be the same age. You would just be counting the years in a different-sized unit.

Exercises—Ages in Mercury and Venusian Years

1. Calculate your age in Mercury years.

_____

_____

2. Venus' year is about 0.62 of an Earth year. What would be your age in Venusian years?

_____

_____

3. If someone were 42 on Earth, what would be his age in Venusian years?

_____

_____

4. If someone were 89 on Earth, what would be her age in Mercury years?

_____

_____

5. If someone were 13 on Earth, what would be his age on Venus?

_____

_____

_____

Name _____ Date _____

Force

In the next section, we will talk about a very important quantity called mass. For some purposes, we will need to relate mass to another important quantity called **force.**

Force can be thought of as a push or a pull on a body. If a force is applied to a body, it will try to change the body's motion—speed it up, slow it down, or change the direction of the motion. If the applied force is opposed by other forces, the motion of the body may not change.

We are used to expressing force in **pounds** (abbreviated **lb**). This is the most common unit used in the United States. Scientists, however, and most of the rest of the world use a unit called a **Newton** (abbreviated **N**) to measure force.

1 Newton = 0.225 pounds
1 pound = 4.45 Newtons

You can convert Newtons to pounds by multiplying the number of Newtons by 0.225 pounds per Newton. For example, 100 Newtons is how many pounds?

100 N x 0.225 lb/N = 22.5 lb

Conversion from pounds to Newtons can be done by multiplying the number of pounds by 4.45 Newtons per pound. For example, 50 pounds is how many Newtons?

50 lb x 4.45 N/lb = 222 N

Exercises—Force

6. How many Newtons is 25 pounds?

_____

7. How many Newtons is 130 pounds?

_____

8. How many pounds is 250 Newtons?

_____

9. How many pounds is 175 Newtons?

_____

10. Weigh yourself on a scale and find out how many pounds you weigh. How many Newtons is this?

_____

Math in Astronomy   Lesson Three—Math: Mass

Name _____ Date _____

Mass

Mass is a measure of how much physical matter is in a body. A body with more matter in it has more mass. It also indicates how much effort (force) is required to speed up, slow down, or change the direction of the body's motion. A football player with more mass is harder to tackle than one with less mass. A car with more mass takes more braking to slow to a stop than one with less mass.

Science uses two mass units, the **gram** (abbreviated **g**) and the **kilogram** (abbreviated **kg**).

1 kilogram = 1,000 grams
1 gram = 0.001 kilograms

You can convert kilograms to grams by multiplying the number of kilograms by 1,000 grams per kilogram. You can convert from grams to kilograms by multiplying the number of grams by 0.001 kilograms per gram.

Exercises—Mass

11. Fifty-five kilograms is how many grams?

_____

_____

12. Twenty kilograms is how many grams?

_____

_____

13. Six thousand grams is how many kilograms?

_____

_____

14. Forty-five grams is how many kilograms?

_____

_____

Math in Astronomy   Lesson Three—Math: Weight

Name _____ Date _____

Weight

Mass is sometimes confused with weight. It is true that the more mass a body has, the more it will weigh on the surface of the earth. **Weight,** however, is usually a measure of how strongly the gravitational attraction of a planet acts on a body. Weight is a force and is measured in Newtons or pounds. If you move to a different planet where the gravitational attraction is different than Earth's, then your weight will be different, but your mass will be the same. If you move out in space somewhere where there is no gravity, your weight will be zero, but your mass will be the same.

Your weight on another planet is equal to your weight on Earth multiplied by the number of Earth gravities that equal that planet's gravitational attraction. For example, suppose you weigh 700 Newtons. If, on a certain fictitious planet, the gravitational attraction was 0.6 Earth gravity, your weight on that planet would be

weight on other planet = 0.6 x 700 N = 420 N

As another example, suppose you weigh 140 pounds. If, on a certain fictitious planet, the gravitational attraction was 2 Earth gravities, your weight on that planet would be

weight on other planet = 2 x 140 lb = 280 lb

Exercises—Weight

15. The gravitational attraction on Mercury is 0.38 of Earth gravity. What would be your weight on Mercury in pounds and in Newtons? Use your weights from Exercise 10.

_____

16. The gravitational attraction on Venus is 0.90 of Earth gravity. What would be your weight on Venus in pounds and in Newtons? Use your weights from Exercise 10.

_____

Converting Weight to Mass and Mass to Weight

You can convert between weight in Newtons **on the surface of the Earth** and mass in kilograms by using the following conversion factors.

To convert from mass in kilograms to weight in Newtons, multiply the number of kilograms by 9.81 Newtons per kilogram. For example, 75 kilograms weighs how many Newtons?

weight = 75 kg x 9.81 N/kg = 736 N

To convert from weight in Newtons to mass in kilograms, multiply the number of Newtons by 0.102 kilograms per Newton. For example, a body that weights 950 Newtons has a mass of how many kilograms?

mass = 950 N x 0.102 kg/N = 96.9 kg

Exercises—Weight and Mass

17. Something that weighs 1,000 Newtons on the surface of the Earth has a mass of how many kilograms?

18. A body that weighs 500 Newtons on the surface of Earth has a mass of how many kilograms?

19. Something that has a mass of 125 kilograms weighs how many Newtons on the surface of Earth?

20. A body that has a mass of 25 kilograms weighs how many Newtons on the surface of Earth?

21. Using your weight in Newtons from Exercise 10, find your mass in kilograms.

Volume

Volume is the word we use to describe how much space something takes up. To occupy volume, a body must be three-dimensional.

To get a feel for volume, let's calculate the volume of probably the simplest three-dimensional object, a box. Mathematicians call this object a **rectangular prism**.

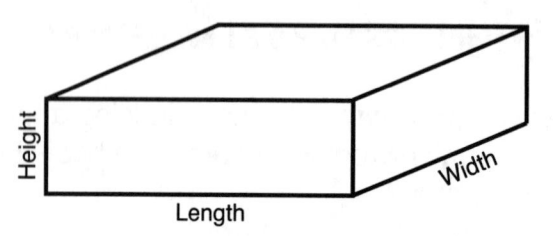

Rectangular Prism

The volume of the prism is just its length times its width times its height. All three of these lengths must be measured in the same units (either centimeters or meters).

volume = length x width x height

Volume is expressed in cubic centimeters (abbreviated **cm³**) or cubic meters (abbreviated **m³**). For example, what is the volume of a box that is 2 meters long, 3 meters wide, and 4 meters high?

volume = 2 m x 3 m x 4 m = 24 m³

As another example, what is the volume of a box that is 5 centimeters long, 3 centimeters wide, and 4 centimeters high?

volume = 5 cm x 3 cm x 4 cm = 60 cm³

Sometimes, we are interested in volumes of other objects, even things as large as planets. Fortunately, we will not have to calculate these volumes in this book.

Math in Astronomy   Lesson Three—Math: Density

Name _____ Date _____

Density

We will use our definitions of mass and volume to introduce a very important property of matter: density. **Density** is a measure of the concentration of matter in a body. If the density is high, there is a lot of mass in a small volume of space. If the density is low, the mass is either small or it is spread out over a large volume of space.

Density is defined as the mass of a body divided by the volume that the body occupies.

$$\text{density} = \frac{\text{mass}}{\text{volume}}$$

The densities of heavenly bodies such as planets are written in grams per cubic centimeter or kilograms per cubic meter.

One gram per cubic centimeter is the same as 1,000 kilograms per cubic meter.

1 g/cm³ = 1,000 kg/m³

You can convert g/cm³ to kg/m³ by multiplying the number of g/cm³ by 1,000 kg/m³ per g/cm³. You can convert from kg/m³ to g/cm³ by multiplying the number of kg/m³ by 0.001 kg/m³ per g/cm³.

Exercises—Densities

22. The density of Venus is 5.24 g/cm³. What is its density in kg/m³?

_____

23. The density of Mercury is 5,400 kg/m³. What is its density in g/cm³?

_____

The densities of planets can be compared to densities of some things with which we are familiar. The density of many rocks on the surface of the Earth is about 3 g/cm³ (3,000 kg/m³). The density of water is 1,000 kg/m³ (1 g/cm³). Anything that is less dense than water will float in it, and anything that is denser than water will sink in it. Very low density materials such as Styrofoam and cork float in water. Denser materials such as rocks or iron sink in water.

The densities of several of the planets are greater than the densities of common rocks. Therefore the materials in the interior of these planets must be very dense. We think that Mercury, Venus, and Earth may have heavy iron materials near their centers.

Temperature: Fahrenheit and Celsius

Temperature is the property that tells us how warm or how cold something is. It is actually a measure of how fast the tiny particles (atoms and molecules) of a substance are moving. The faster they are moving, the higher the temperature.

We are used to thinking of temperature in degrees. A degree is just a part of a temperature scale. The temperature scale that most of us know best is the Fahrenheit scale. On this scale, the temperature at which water freezes is given the value 32 degrees, and the temperature at which water boils is given the value 212 degrees. One hundred eighty divisions on the scale separate these two values. These divisions define the size of a degree. Fahrenheit temperature is usually abbreviated as °F.

There are other temperature scales in use. One of them is the Celsius scale. On this scale, the freezing temperature of water is at zero degrees, and the boiling temperature of water is 100 degrees. Celsius temperature is usually abbreviated as °C.

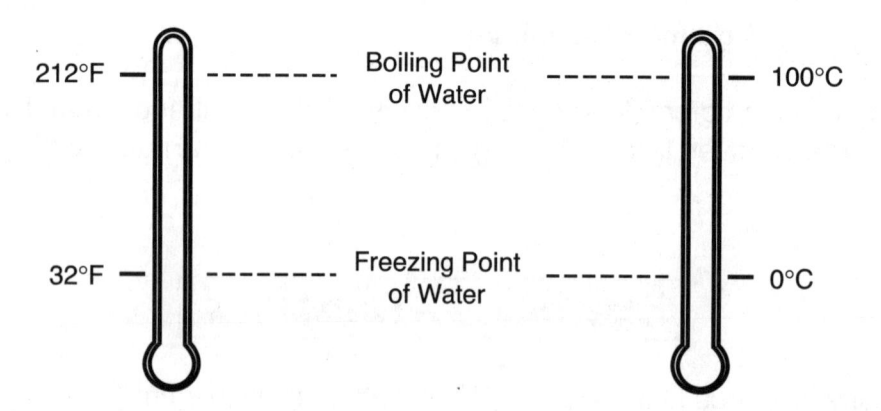

Fahrenheit Thermometer — 212°F Boiling Point of Water — 100°C **Celsius Thermometer**

32°F Freezing Point of Water — 0°C

You can convert between these two temperature scales by using the following conversion equations.

$$\text{Fahrenheit temperature} = \frac{9}{5} \times \text{Celsius temperature} + 32$$

$$\text{Celsius temperature} = \frac{5}{9} \times (\text{Fahrenheit temperature} - 32)$$

A temperature of 15° C is how many degrees Fahrenheit?

$$\text{Fahrenheit temperature} = \frac{9}{5} \times 15 + 32 = 9 \times 3 + 32$$

$$= 27 + 32 = 59° F$$

A temperature of 50° F is how many degrees C?

$$\text{Celsius temperature} = \frac{5}{9} \times (50 - 32) = \frac{5}{9} \times 18$$

$$= 5 \times 2 = 10° C$$

Name _____ Date _____

Exercises—Temperature: Fahrenheit and Celsius

24. A temperature of 100° C is how many degrees Fahrenheit?

_____

25. A temperature of 25° C is how many degrees Fahrenheit?

_____

26. A temperature of -5° C is how many degrees Fahrenheit?

_____

27. A temperature of 95° F is how many degrees Celsius?

_____

28. A temperature of 77° F is how many degrees Celsius?

_____

29. A temperature of -13° F is how many degrees Celsius?

_____

Temperature: Kelvin

A temperature scale that is often used in science is called the Kelvin scale. The Kelvin scale has a zero point that is at the lowest possible temperature that anything can have. This temperature is about 273° below zero on the Celsius scale. Temperature on the Kelvin scale is measured in Kelvins (abbreviated K) instead of degrees.

The conversion equations below show how to convert between temperature on the Celsius scale and temperature on the Kelvin scale.

Kelvin temperature = Celsius temperature + 273
Celsius temperature = Kelvin temperature - 273

For example, a temperature of 30° C would be how many Kelvins?

Kelvin temperature = 30 + 273 = 303 Kelvins

Exercises—Temperature: Kelvin

30. A temperature of 45° C is how many Kelvins?

_____

31. A temperature of 450 K is how many degrees Celsius?

Math in Astronomy   Lesson Three—Teacher's Page

Lesson Three—Teacher's Page

Class Activities

I. Ask the students to list the weights of various family members in pounds. Convert these weights to Newtons. Find the masses of these family members.

II. Ask the students to speculate about what their weights would be if they were very far from any planet, so that the gravitational attraction of any planet was effectively zero. Would they be able to walk around upright on a surface?

Do the students "feel weightless" when they jump off of a high diving board during the time that they are in the air? Are you "weightless" when you are falling? The astronauts appear "weightless" on the space shuttle. Are they falling around the earth in their orbit?

Why do the characters on *Star Trek* appear to "feel" gravity? Is there some sort of fictitious artificial gravity generator?

III. Find or cut some cubes of common materials, some of which float and some of which sink. Have the students measure their dimensions and compute their volumes. Then have them obtain the masses of the cubes and compute their densities. Ask them to predict which cubes will sink or float. Then, drop them in a tub of water and see which ones sink or float.

IV. Find a good Fahrenheit thermometer. Ask your students to measure the temperature in and around your classroom. Find a fever thermometer and, after sterilizing it with alcohol, measure a student's body temperature. Have the students convert these numbers to Celsius and then to Kelvin.

Answers to Questions (page 36)

1. b   2. d   3. a   4. a   5. d   6. b   7. a

Answers to Mathematical Exercises (pages 37–45)

3. 67.74 yrs	4. 370.83 yrs	5. 20.97 yrs	
6. 111 N	7. 579 N	8. 56 lb	9. 39 lb
11. 55,000 g	12. 20,000 g	13. 6 kg	14. 0.045 kg
17. 102 kg	18. 51 kg	19. 1226 N	20. 245 N
22. 5,240 kg/m^3	23. 5.4 g/cm^3		
24. 212° F	25. 77° F	26. 21° F	27. 35° C
28. 25° C	29. -25° C		
30. 318 K	31. 177° C		

Lesson Four
The Earth

Introduction

Of all the planets in the solar system, we know the most about Earth. It is, as far as we know, the only planet in the solar system that supports life. For all we know, it may be the only planet in the universe that can support life, although we have some reasons to suspect that this may not be true.

Earth is unlike Venus. It does not have a thick oppressive atmosphere, and its surface temperature is not unbearably hot. Clouds often form in Earth's atmosphere, but they never completely cover the planet. They are composed of water vapor, not nasty chemicals such as sulfuric acid. Also, unlike Venus, Earth has a lot of water available on its surface.

As we will see in Lesson Six, Earth is also unlike Mars, the next planet outward in the solar system.

Earth Properties

Earth orbits the Sun at a comfortable distance of one Astronomical Unit. Its orbit is not very eccentric, so its path around the Sun is almost circular.

Earth Properties

Average distance from the Sun	1.00 AU
Orbit period	1.00 years
Orbit eccentricity	0.017
Inclination of equator to orbit	23.5°
Sidereal rotation period	23 hours 56.07 min
Diameter	12,760 km
Mass	5.97×10^{24} km
Average density	5,520 kg/m^3

The Seasons

We owe our seasons, Spring, Summer, Fall, and Winter, to the revolution of the earth around the Sun and to the 23.5° tilt of its axis. When Earth is in a certain position in its orbit, the tilt of its axis causes the Northern Hemisphere to receive the Sun's rays at a more direct angle than it does during the other times of the year.

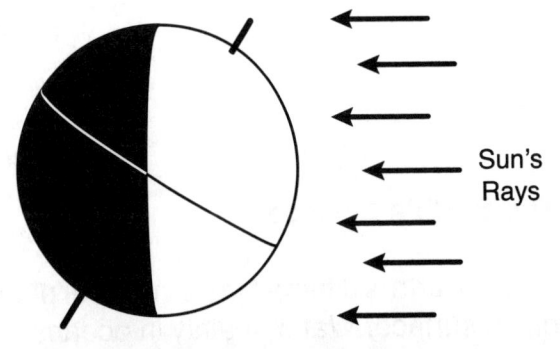

The Sun's Rays During Summer in the Northern Hemisphere

This is the time of the year that we (in the Northern Hemisphere) call Summer. Not only do we get the Sun's rays more directly, but we have longer days as well. The Sun rises earlier in the morning and sets later in the evening than at other times of the year. The day with the most daylight is called the **summer solstice.** It usually occurs on June 21 and is called the first day of Summer.

While we are having Summer in the Northern Hemisphere, the people in the Southern Hemisphere are experiencing Winter.

When the earth is on the other side of its orbit, we experience Winter in the Northern Hemisphere.

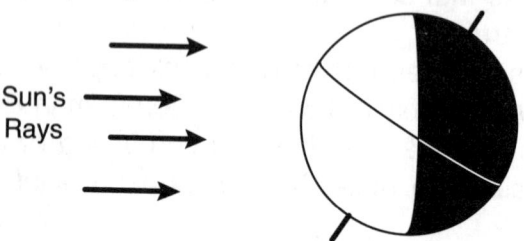

The Sun's Rays During Winter in the Northern Hemisphere

The Sun's rays fall on the earth at a smaller angle than at other times of the year, and the daylight portions of days are shorter. The day with the least Sunlight is called the **winter solstice.** It falls on December 21, and it is the official first day of Winter.

The orbit positions of the earth about halfway between Summer and Winter are called **equinoxes.** There is one that occurs on March 21 and another one that falls on September 23. The one in March is called the **vernal equinox** and is the first day of Spring. The one in September, called the **autumnal equinox,** begins the Fall season. On an equinox day, every place on Earth experiences twelve hours of daylight and twelve hours of darkness.

The picture below shows the relative positions of the Summer and Winter solstices and the Vernal and Autumnal equinoxes.

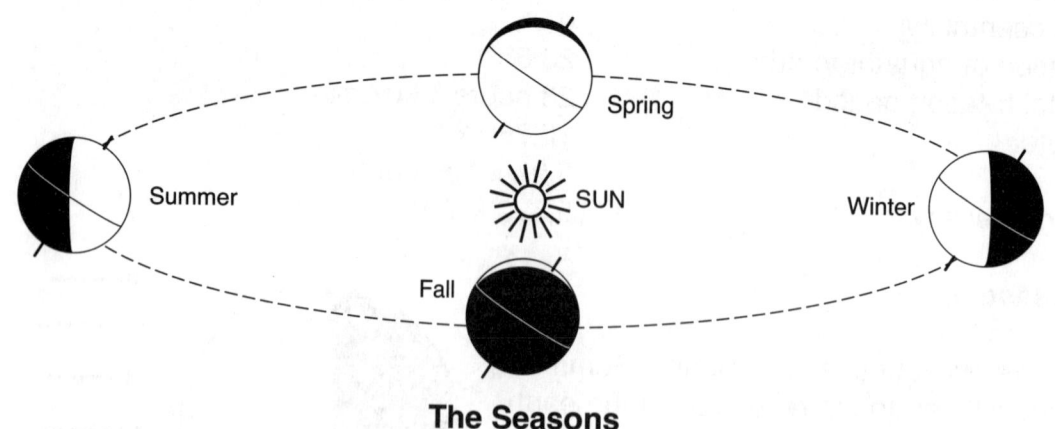

The Seasons

The Earth's Surface

Earth is different from the other planets in the solar system because it has liquid water on its surface. Water, mainly in oceans and seas, covers about 70 percent of the surface of the earth. Areas of land that rise above the water are called continents if they are large and islands if they are small. There are seven continents on the earth and a very large number of islands.

Scientists have found that many of the landforms on the earth's surface can be explained by a theory called **plate tectonics.** There is a lot of evidence to support this theory. Plate tectonics says that the surface of the earth is divided into several large plates of hard rock that float on top of a pliant layer of rock underneath. These plates can move with respect to one another. They do not move rapidly (only a couple of centimeters per year), but they do move.

The picture below shows the continents in bold outline and Earth's main plates in lighter outline.

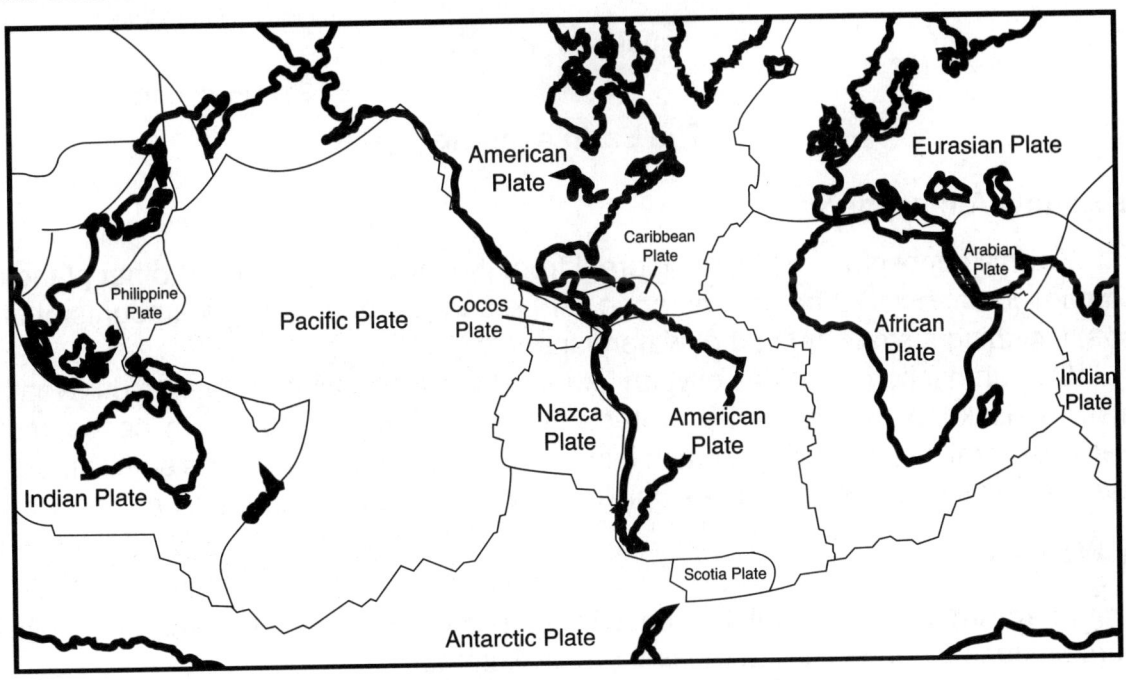

The Main Plates on Earth's Surface

Interesting things can happen at the boundaries of these plates. In the places where two plates are moving away from each other (as they are in the middle of the Atlantic Ocean), liquid rock can ooze up through the crack between the plates. We see this happening in underwater volcanoes on the floor of the Atlantic Ocean.

In other places, where two plates are moving toward each other, one plate can dive under the other one. When this happens on the floor of an ocean, it forms a very deep ocean trench. We think that this process has formed the Mariana Trench (over 11 km deep) in the western Pacific Ocean. In yet other places, land is forced upward when two plates move toward one another. We see this in Asia where the collision of two plates has formed the Himalaya Mountains.

Regions along the edges of the plates are places where volcanos and earthquakes are common.

The Earth's Interior

The interior of the earth is much denser and much different than what we see on the surface. The surface rocks that we see have an average density of about 3 g/cm^3, but the average density of the whole earth is much higher—about 5.52 g/cm^3.

Scientists are pretty sure that there is a dense **core** in the center of the earth that is composed mostly of iron. The innermost portion of this core is solid, but the outer part is molten (liquid). A less dense, rocky region, called the **mantle,** surrounds the core. A thin, low-density layer called the **crust** covers the surface of the earth.

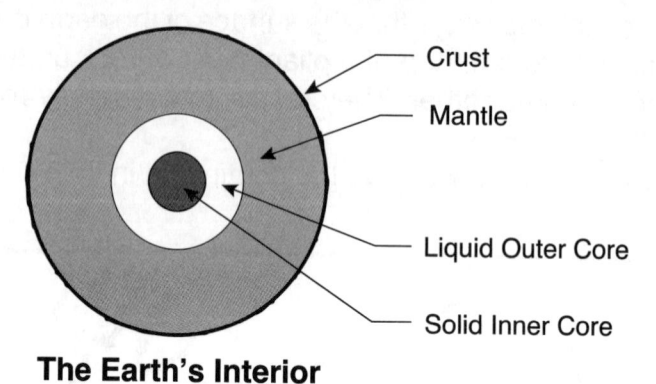

The Earth's Interior

The Earth's Atmosphere

Earth's atmosphere is much different from the atmospheres of the other planets. It is composed of about 78 percent nitrogen, 21 percent oxygen, and tiny amounts of other gases, including a small amount of water vapor.

Our atmosphere provides oxygen for us to breath and wind to change our weather. It also helps keep us warm and protects us from falling bodies from space (called meteoroids) that might otherwise crash into the surface of our planet. The small amount of water vapor in the atmosphere is enough to allow clouds to form and rain and snow to fall.

Key Words

Summer solstice: The day with the most daylight (in the Northern Hemisphere). It usually occurs on June 21 and is called the first day of Summer.

Winter solstice: The day with the least amount of daylight (in the Northern Hemisphere). It falls on about December 21 and is the official first day of Winter.

Equinox: Day on which there are twelve hours of daylight and twelve hours of darkness everywhere on Earth.

Vernal equinox: An equinox that is the first day of the Spring season (in the Northern Hemisphere). It falls on about March 21.

Autumnal equinox: An equinox that is the first day of the Fall season (in the Northern Hemisphere). It usually occurs on September 23.

Plate tectonics: The theory that says that the surface of the earth is divided into several large plates of hard rock that float on top of a plastic layer of rock underneath.

Core: The center portion of the earth. It is composed mostly of iron. The inner portion of the core is solid, but the outer portion is liquid.

Mantle: The interior region of the earth between the core and the crust. It is not as dense as the core, but it is denser than the crust.

Crust: The outer portion of the earth, near and at the surface. Its density is low.

Math in Astronomy

Lesson Four—Questions

Name _____ Date _____

Lesson Four—Questions

1. In the Northern Hemisphere, the day with the most daylight is called the _____. This day usually occurs on June 21.

    a) Summer solstice   b) Winter solstice   c) Vernal equinox   d) Autumnal equinox

2. In the Northern Hemisphere, the day with the least daylight is called the _____. This day usually occurs on December 21.

    a) Summer solstice   b) Winter solstice   c) Vernal equinox   d) Autumnal equinox

3. A day on which there are twelve hours of darkness and twelve hours of daylight everywhere on Earth is called a(n)

    a) solstice   b) Winter solstice   c) Summer solstice   d) equinox

4. An equinox that is the first day of Spring (in the Northern Hemisphere) is called the _____. It usually occurs on March 21.

    a) Summer solstice   b) Winter solstice   c) Vernal equinox   d) Autumnal equinox

5. An equinox that is the first day of Autumn (in the Northern Hemisphere) is called the _____. It usually occurs on September 23.

    a) Summer solstice   b) Winter solstice   c) Vernal equinox   d) Autumnal equinox

6. The semimajor axis of Earth's orbit is 1.00 AU, and the semi-major axis of Venus's orbit is 0.72 AU. How far is Earth from Venus when they are in the position shown below?

    a) 1.72 AU   b) 1.28 AU   c) 0.28 AU   d) 0.72 AU

© Mark Twain Media, Inc., Publishers

Lesson Four—Questions (continued)

7. The semimajor axis of Earth's orbit is 1.00 AU, and the semi-major axis of Mars' orbit is 1.52 AU. How far is Earth from Mars when they are in the positions shown below?

   a) 1.52 AU b) 0.52 AU c) 2.52 AU d) 0.48 AU

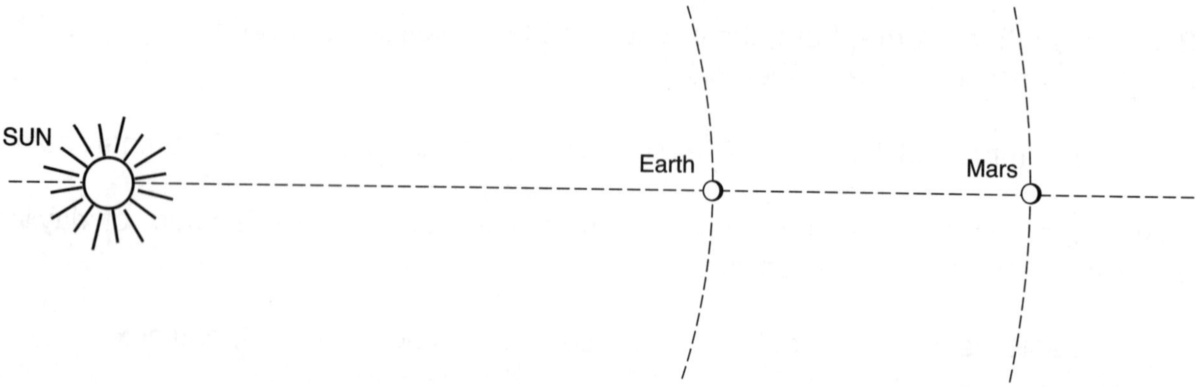

8. The semimajor axis of Earth's orbit is 1.00 AU, and the eccentricity of its orbit is 0.017. What is Earth's perihelion distance?

   a) 0.98 AU b) 1.02 AU c) 0.00 AU d) 0.034 AU

9. Using the data from question 8, what is Earth's aphelion distance?

   a) 0.98 AU b) 1.02 AU c) 0.00 AU d) 0.034 AU

10. The center portion of the earth, composed mostly of iron is called the

    a) crust b) mantle c) heart d) core

11. The outer portion of the earth, near and at the surface that has a low density is called the

    a) crust b) mantle c) heart d) core

12. The theory that says the earth's surface is divided into several large plates of hard rock that float on top of a plastic layer of rock underneath is called

    a) plate theory
    c) plate displacement
    b) plate tectonics
    d) plate drift

Name _____ Date _____

Lesson Four—Math

Hours and Minutes

We are used to describing time in hours and minutes (and sometimes seconds, too). We know that there are 60 minutes in an hour. For some purposes in astronomy, we need to convert some number of hours to a number of minutes, or to convert some number of minutes to hours. Fortunately, these conversions are easy.

To convert hours to minutes, simply multiply the number of hours by 60 min/hr. For example, convert 5 hours to minutes.

5 hrs x 60 min/hr = 300 min

In some conversions, you may end up with a fraction of a minute. For our purposes in this book, we will round off our answers to the nearest minute. As an example, convert 9.34 hours to minutes.

9.34 hrs x 60 min/hr = 560.4 min
~ 560 min

Exercises—Hours to Minutes Conversion

Round to the nearest minutes.

1. Convert 3 hours to minutes. _____

2. Convert 11 hours to minutes. _____

3. Convert 6.25 hours to minutes. _____

4. Convert 8.71 hours to minutes. _____

Name _____ Date _____

Hours and Minutes (continued)

To convert from minutes to hours, you just need to divide by 60.
   For example, convert 150 minutes to hours.

150 min ÷ 60 min/hr = 2.5 hrs

Exercises—Minutes to Hours Conversion

5. Convert 540 minutes to hours. _____

6. Convert 270 minutes to hours. _____

7. Convert 435 minutes to hours. _____

8. Convert 187 minutes to hours. _____

   For some purposes, you may want to convert a large number of minutes to hours with some minutes (less than 60) left over.
   To do this, divide the number of minutes by 60 min/hr in a way that gets a quotient and a remainder. Then add the quotient to the hours and use the remainder as the minutes.
   For example, convert 135 minutes to hours and minutes.

**135 min ÷ 60 min/hr = 2 with a remainder of 15
= 2 hr 15 min**

Exercises—Minutes to Hours and Minutes Conversion

9. Convert 247 minutes to hours and minutes. _____

10. Convert 225 minutes to hours and minutes. _____

11. Convert 575 minutes to hours and minutes. _____

12. Convert 359 minutes to hours and minutes. _____

Math in Astronomy   Lesson Four—Math: The Twenty-Four-Hour Clock

Name _____ Date _____

The Twenty-Four-Hour Clock

Most of us use a clock that shows twelve hours. We have twelve hours in the morning that we call A.M. time, and twelve more hours in the afternoon and evening that we call P.M. time. We know that 2:30 A.M. is in the wee hours of the morning and that 2:30 P.M. is in the afternoon.

In astronomy, we often use a 24-hour clock. A 24-hour clock does not have A.M. and P.M. time. It just has one time that we call **hours.** Zero (00) hours is at midnight, and noon occurs at a time of 12 hours. A time of 24 hours is midnight again, the same as 00 hours. This is also the kind of time used in the military that many of you may have heard in movies or on television programs.

We can, of course, express our time as hours and minutes. Sometimes, it is written as **number of hours:number of minutes** followed by the word **hours.** For example, 07:30 hours or 14:15 hours. In astronomy, we often write the number of hours, followed by the letter, **h,** and then the number of minutes, followed by the letter, **m.** For examples, 5h 25m and 13h 20m.

A.M. times are the same on a 24-hour clock as they are on an ordinary 12-hour clock. P.M. times, however, are 12 hours larger on a 24-hour clock than they are on a 12-hour clock.

To convert a P.M. time to a 24-hour clock time, add 12 hours.
For example, convert 5:30 P.M. to a 24-hour clock time.

24-hour time = 5h 30m + 12h
                = 17h 30 m

To convert a 24-hour clock time to a P.M. time, subtract 12 hours.
For example, convert 21h 15m to P.M. time.

P.M. time = 21h 15m - 12h
           =   9h 15m

Exercises—Twenty-Four-Hour Clock

13. What is 6:45 P.M. on a 24-hour clock?

14. What is 11:40 P.M. on a 24-hour clock?

15. What is 10:20 A.M. on a 24-hour clock?

16. Convert 19h 40m to A.M. or P.M. time on a 12-hour clock.

17. Convert 16h 55m to A.M. or P.M. time on a 12-hour clock.

18. Convert 9h 25m to A.M. or P.M. time on a 12-hour clock.

Adding and Subtracting Times

Sometimes (in the next section of this lesson, for example) we have to add and subtract times that are expressed in hours and minutes. Sometimes this is easy, but sometimes it has a few surprises. In the simplest case, you add two times by just adding the hours and then adding the minutes.

For example, add 10h 22m and 5h 18m.

```
   10h 22m
+   5h 18m
   15h 40m
```

Suppose, however, that the sum of our two times gives you more than 60 minutes. Then you have to subtract 60 from your minutes sum and add 1 to your hours sum.

For example, find 9h 45m + 6h 35m.

```
    9h 45m
+   6h 35m
   15h 80m = 16h 20m
```

There is one more problem that you might encounter when you add times. Suppose that your sum is greater than 24 hours. When this happens, your time sum has extended into the next day. You must subtract 24 hours from your sum.

For example, find 13h 38m + 15h 10m.

```
   13h 38m
+  15h 10m
   28h 48m
-  24h
    4h 48m
```

Subtraction is very simple (if you don't have to "borrow").

For example, subtract 7h 26m from 13h 42m.

```
   13h 42m
-   7h 26m
    6h 16m
```

Name _____ Date _____

Adding and Subtracting Times (continued)

If the number of minutes that you subtract, however, is more than the number of minutes that you subtract from, you have to borrow 60 minutes from the hour of the time you subtract from.

For example, find 11h 12m - 6h 30m.

$$\begin{array}{r} 11h\ 12m \\ -\ \ 6h\ 30m \end{array} \Rightarrow \begin{array}{r} 10h\ 72m \\ -\ \ 6h\ 30m \\ \hline 4h\ 42m \end{array}$$

Finally, there is another problem that you might run into when subtracting times. You may get a negative time as your result. Your subtraction has taken your time back to the previous day. When this happens, add 23h 60m to your result. You will get the correct time on the previous day. If only the minutes are negative, this may put you over 24 hours, so remember to subtract 24 from the hours if that happens.

For example, find 9h 35m - 12h 45m

$$\begin{array}{r} 9h\ \ 30m \\ -\ 12h\ \ 45m \\ \hline -3h\ -15m \end{array} \Rightarrow \begin{array}{r} -3h\ -15m \\ +\ 23h\ \ 60m \\ \hline 20h\ \ 45m \end{array}$$

Exercises—Adding and Subtracting Times

19. Find the following sums.
    a) 7h 14m
       + 5h 19m

    b) 15h 29m
       + 2h 46m

    c) 12h 41m
       +10h 39m

    d) 0h 22m
       +8h 18m

    e) 13h 09m
       +15h 21m

    f) 14h 45m
       +21h 55m

20. Find the following differences.
    a) 22h 55m
       - 9h 33m

    b) 14h 23m
       -10h 23m

    c) 11h 28m
       -19h 58m

    d) 16h 15m
       - 9h 45m

    e) 8h 30m
       -3h 50m

    f) 22h 12m
       -14h 57m

Sidereal and Solar Days

Our Earth rotates on its axis and also moves in its orbit around the Sun. Because it has these two motions, we have two kinds of days—with different lengths! The first kind, a **sidereal day,** is the simplest. A sidereal day is the time it takes the earth to rotate around once with respect to the stars. Suppose you stuck a stick in the ground at night so that it pointed to a star. The next night, after the earth had rotated once, that stick would point to the same star. The time that passed would be a sidereal day.

The other kind of day, a **solar day,** is a little longer. It is the time it takes the earth to rotate around once with respect to the Sun. Suppose you stuck a stick in the ground at noon so that it pointed to the Sun. The next day, again at noon, the stick would again point at the Sun. One solar day would have passed. To see why the solar day is a little longer, look at the diagram below.

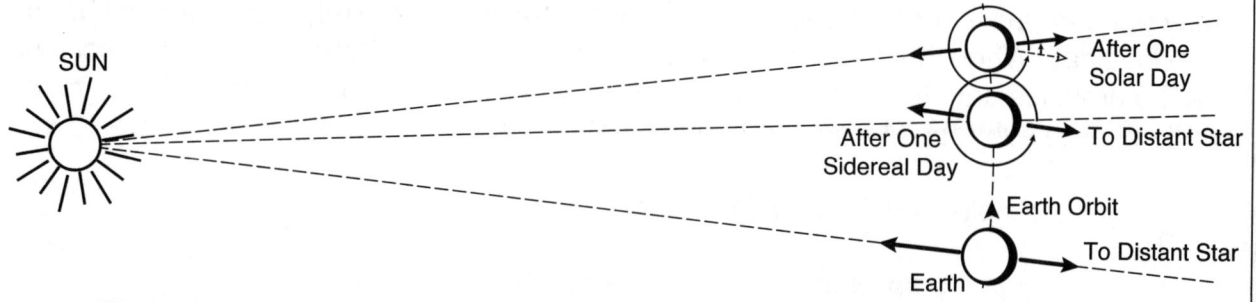

The stick in the earth is shown as an arrow in the diagram. To start with, it points at the Sun. After the earth rotates once, with respect to the stars, the arrow points in the same directions that it pointed when the rotation started. The earth, however, has moved in its orbit so that the arrow no longer points at the Sun. The earth has to rotate a little more to realign the pointer with the Sun. Therefore, a solar day is just a little longer than a sidereal day. (The angles and distances in the diagram are exaggerated.)

We define a solar day to be 24 hours. A sidereal day turns out to be 23 hours and 56.07 minutes. Our clocks run on solar time, because we want the Sun to be highest in the sky at about noon each day. This means that the stars will rise about 4 minutes (actually 3.93 minutes) earlier each night, with respect to our solar time clocks. We notice this if we watch the sky each night for several months. Stars that appeared in the eastern sky at a certain time in the evening one month will be seen higher in the sky, overhead, or even in the western sky at the same time the next month.

If we know the time that a particular star rises above the eastern horizon on a certain date, we can calculate the (approximate) time that star would rise on a later date.

time star rises tonight - (4m x number of days until later date) = time star rises on later date

For example, suppose that a certain star rises at 10:30 tonight. What time would it rise four nights from now?

20h 30m - (4m x 4 days) =
20h 30m - 16m =
20h 14m = time four nights from now

Name _____ Date _____

Exercises—Star Rise Times

21. Suppose a certain star rises at 23h 55m tonight. What time will it rise ten nights from now?

_____

22. Suppose a certain star rises at 21h 32m tonight. What time will it rise five nights from now?

_____

23. Suppose a certain star rises at 22h 10m tonight. What time will it rise 12 nights from now?

_____

24. Suppose a certain star rises at 21h 12m tonight. What time will it rise 20 nights from now?

_____

25. Suppose a certain star rises at 10:40 P.M. on October 10. What time will it rise on October 17?

_____

26. A certain star rises at 11:35 P.M. on March 29. What time will it rise on April 3?

_____

27. Suppose a certain star rises at 01h 18m tonight. What time will it rise three nights from now?

_____

28. Suppose a certain star rises at 00h 14m tonight. What time will it rise ten nights from now?

_____

29. Suppose that you know the time that a star rises tonight. Figure out a way of calculating when that star would have risen last night, the night before, or several nights ago.

_____

Latitude and Longitude

Most of us have seen or used maps at some time in our lives. We may wonder how the maps were made and how the person who drew them knew where to place the towns, rivers, and boundaries. The answer is that he or she used coordinates (numbers) to describe how far north or south something was and how far east or west it was.

We live on a spherical planet. If we want to describe our position at any place on this sphere, we need some sort of coordinate system that can pinpoint any spot on the globe. The coordinates that we use are called latitude and longitude. **Latitude** describes how far north or south of the earth's equator a position is, and **longitude** describes how far east or west of some reference point it is.

Latitude is the angle between a spot on the earth and the equator, as seen from the center of the earth. It is measured in degrees.

Latitudes can range from 0° at the equator to 90° at the North Pole or South Pole. Latitude is called north latitude if it applies to a place in the earth's northern hemisphere and south latitude if it refers to a place in the earth's southern hemisphere. We sometimes draw lines on maps and models of the earth to represent certain latitudes. These are called **parallels** of latitude. They are circles around the earth that are parallel to the equator.

In order to understand longitude, it is necessary to talk about something called a **meridian**. A meridian is a circle drawn on the earth's surface that passes through the North Pole and the South Pole.

There is one meridian that is special. It is called the **prime meridian**, and it passes through Greenwich, England. We define the longitude of some spot on the earth as the angle between a meridian passing through that spot and the prime meridian. Like latitude, longitude is measured in degrees.

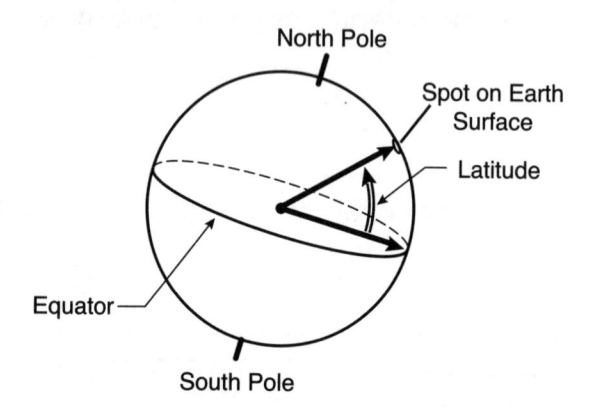

Latitude of a Point on the Earth

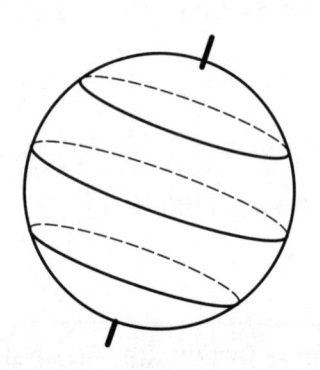

Parallels of Latitude on the Earth

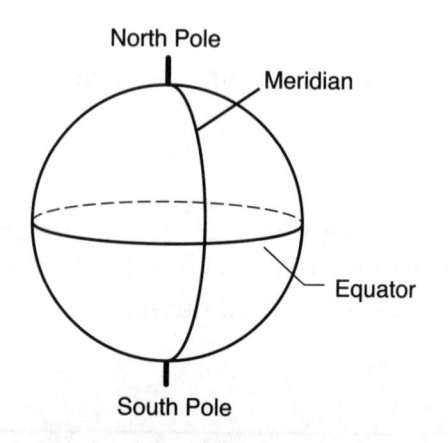

A Meridian of Longitude

Latitude and Longitude (continued)

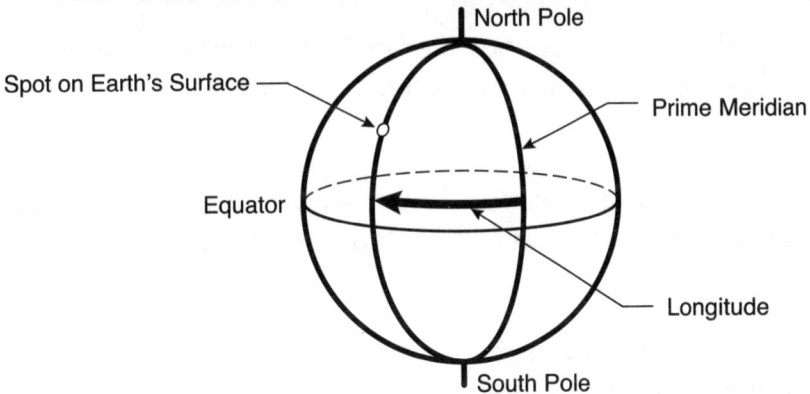

Longitude of a Point on the Earth

A longitude can be a number between 0 and 180°. Places to the east of the prime meridian have east longitudes, and places to the west of the prime meridian have west longitudes.

Since longitude only goes up to 180°, there should be a meridian on the earth, opposite the prime meridian, that is at both 180° east longitude and 180° west longitude. Such a meridian exists near the center of the Pacific Ocean.

Exercises—Latitude and Longitude

30. Your teacher will provide you with a globe and one or more maps. Find the latitudes and longitudes of the places on the globe and maps that he or she suggests (to the nearest degree.)

31. Find the latitude and longitude of your school (to the nearest degree).

Elevation of the Sun

If you have glanced at the Sun during different times of the year, you have probably noticed that it seems higher in the sky during Summer and lower in the sky during Winter. This is indeed true.

NOTE: Looking directly at the Sun, even with sunglasses on, is extremely dangerous and can lead to eye damage and blindness. Do not look at the Sun.

We sometimes measure how high the Sun is in the sky by measuring the angle between it and the southern horizon at noon. We call this angle the **elevation angle** of the Sun. It changes with the seasons, and it depends on the latitude of the observer. It is easy to calculate, however, for the Summer Solstice, the Winter Solstice, and the Vernal and Autumnal Equinoxes. We will do these calculations for observers in the Northern Hemisphere of the earth.

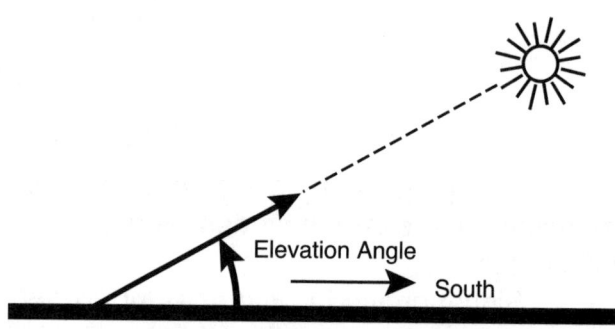

Elevation Angle of the Sun

At the Summer Solstice

On the Summer Solstice, the Sun appears directly overhead at noon to an observer at a latitude of 23.5°N. To observers at other latitudes in the Northern Hemisphere, it would be at the elevation angle given below.

elevation angle = 90 - (latitude - 23.5)
   = 90 + 23.5 - latitude
   = 113.5 - latitude

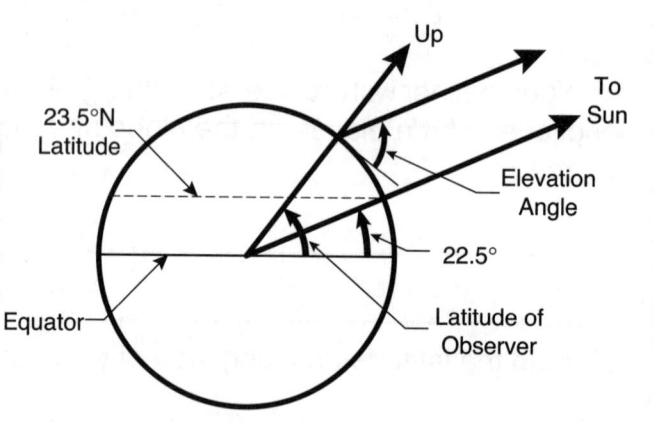

Elevation Angle of the Sun at the Summer Solstice

At the Winter Solstice

On the Winter Solstice, the Sun appears directly overhead at noon to an observer at a latitude of 23.5°S. To observers at latitudes in the Northern Hemisphere, it would be at the elevation angle given below.

elevation angle = 90 - (23.5 + latitude)
   = 90 - 23.5 - latitude
   = 66.5 - latitude

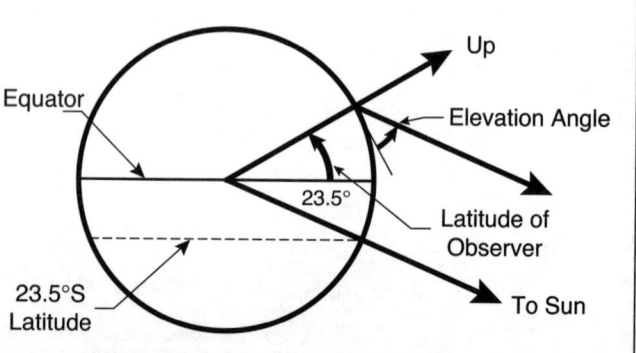

Elevation Angle of the Sun at the Winter Solstice

Math in Astronomy   Lesson Four—Math: Elevation of the Sun

Name _____ Date _____

At the Equinoxes

On the Vernal and Autumnal Equinoxes, the Sun appears directly overhead to an observer at the equator. To observers at latitudes in the Northern Hemisphere, it would be at the elevation angle given below.

elevation angle = 90 - latitude

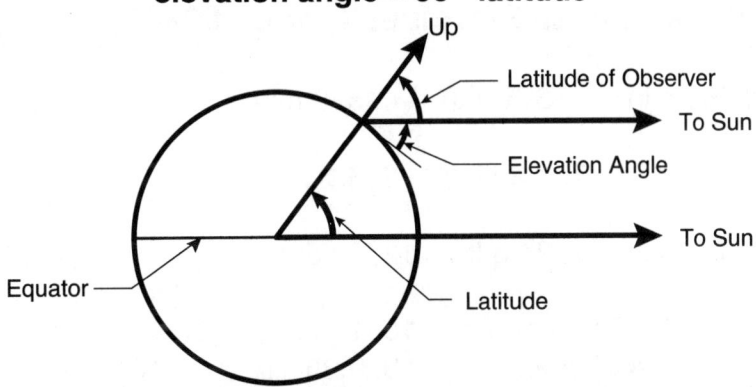

Elevation Angle of the Sun at the Equinoxes

Exercises—Elevation of the Sun

32. For a latitude of 30°N, compute the elevation of the Sun on
a) the Summer Solstice; _____
b) the Winter Solstice; _____
c) the Vernal or Autumnal Equinox. _____

33. For a latitude of 45°N, compute the elevation of the Sun on
a) the Summer Solstice; _____
b) the Winter Solstice; _____
c) the Vernal or Autumnal Equinox. _____

34. For a latitude of 60°N, compute the elevation of the Sun on
a) the Summer Solstice; _____
b) the Winter Solstice; _____
c) the Vernal or Autumnal Equinox. _____

35. Compute the elevation of the Sun for your latitude on
a) the Summer Solstice; _____
b) the Winter Solstice; _____
c) the Vernal or Autumnal Equinox. _____

© Mark Twain Media, Inc., Publishers

Lesson Four—Teacher's Page

Answers to Questions (pages 51–52)

1. a 2. b 3. d 4. c 5. d 6. c 7. b 8. a 9. b 10. d 11. a 12. b

Answers to Mathematical Exercises (pages 53–63)

1. 180 min   2. 660 min   3. 375 min   4. 523 min

5. 9 hr   6. 4.5 hr   7. 7.25 hr   8. 3.1 hr

9. 4h 7m   10. 3h 45m   11. 9h 35m   12. 5h 59m

13. 18h 45m   14. 23h 40m   15. 10h 20m
16. 7:40 P.M.   17. 4:55 P.M.   18. 9:25 A.M.

19. a) 12h 33m;   b) 18h 15m;   c) 23h 20m;   d) 8h 40m;
    e) 4h 30m (next day);   f) 12h 40m (next day)
20. a) 13h 22m;   b) 4h 0m;   c) 15h 30m (previous day);   d) 6h 30m;
    e) 4h 40m;   f) 7h 15m

21. 23h 15m   22. 21h 12m   23. 21h 22m   24. 19h 52m
25. 22h 12m or 10:12 P.M.   26. 23h 15m or 11:15 P.M.
27. 01h 06m   28. 23h 34m
29. Students should arrive at a formula such as:

$$\text{time star rises tonight} + (4m \times \text{number of days to earlier date}) = \text{time star rises on earlier date}$$

32. a) 83.5°;
    b) 36.5°;
    c) 60°
33. a) 68.5°;
    b) 21.5°;
    c) 45°
34. a) 53.5°;
    b) 6.5°;
    c) 30°
35. Teacher check answers for your latitude.

Lesson Five
The Moon

Introduction

The Moon is our closest neighbor in space and, for time unknown, it has held a special attraction for people. When we were young, we heard nursery rhymes about the cow jumping over the Moon and tales about the "man in the Moon." Even before spaceflight was possible, authors like Jules Verne wrote stories about travel to the Moon.

The Moon is close enough that it can be viewed with binoculars or small telescopes. We can see the dark and light areas on its surface that led early observers to believe that there were seas and land masses there. Scientists have observed the Moon with large telescopes for a long time and have learned a lot from their observations.

After July 20, 1969, the Moon had an even greater appeal. That was the date that an astronaut first stepped onto its surface. Because of that visit and five others, we now have rocks and a lot of scientific data from the Moon. Our last visit to the Moon was in 1972, but we are still learning things from the rocks and the data.

The Moon looks big to us because it is close and because it _is_ big. Its diameter is 0.27 of Earth's diameter and its mass is 0.012 Earth masses. That is bigger than the planet Pluto. Some scientists have called Earth and its moon a double planet.

The Moon's Orbit

The Moon travels around the Earth with a sidereal period of about 27.3 days. Its rotation period is also 27.3 days. Therefore, it takes the same amount of time to complete a rotation that it takes to complete a revolution. This means that any time we look at the Moon we always see the same side. The other side was not revealed to us until satellites were placed in orbit around the Moon and directed to photograph it.

The lunar orbit is not a perfect circle. It is an ellipse with a semi-major axis of about 384,000 km (0.00257 AU) and an eccentricity of 0.055. The plane of its orbit is tilted at an angle of about 5° to the plane of Earth's orbit.

The Moon's Phases

When we look at the Moon, we do not always see the same shape. Sometimes it looks like a crescent, sometimes it looks like a half circle, and at other times it looks like a full disk. We say that the moon goes through **phases.** A **phase** is just an appearance or shape of the sunlit portion of the Moon as seen from Earth.

The reason for the Moon's phases is easy to understand. At all times, half of the Moon's surface (the face toward the Sun) is sunlit. As the Moon travels around the earth in its orbit, we see the sunlit face, or portions of it, from different angles. When the Moon is between us and the Sun, we can only see the dark side that is not sunlit. (It is also difficult to view the Moon when it is in the same direction as the Sun.) When this happens, we say that we have a **new moon.** When the Moon is on the opposite side of the earth from the Sun, we see it high in the night sky, and it looks like a bright disk. We say we have a **full moon.**

Other phases are called **waxing crescent, first quarter, waxing gibbous, waning gibbous, last quarter,** and **waning crescent.** The diagram below shows the orbit positions of the moon and the phases that we see when the Moon is in those positions.

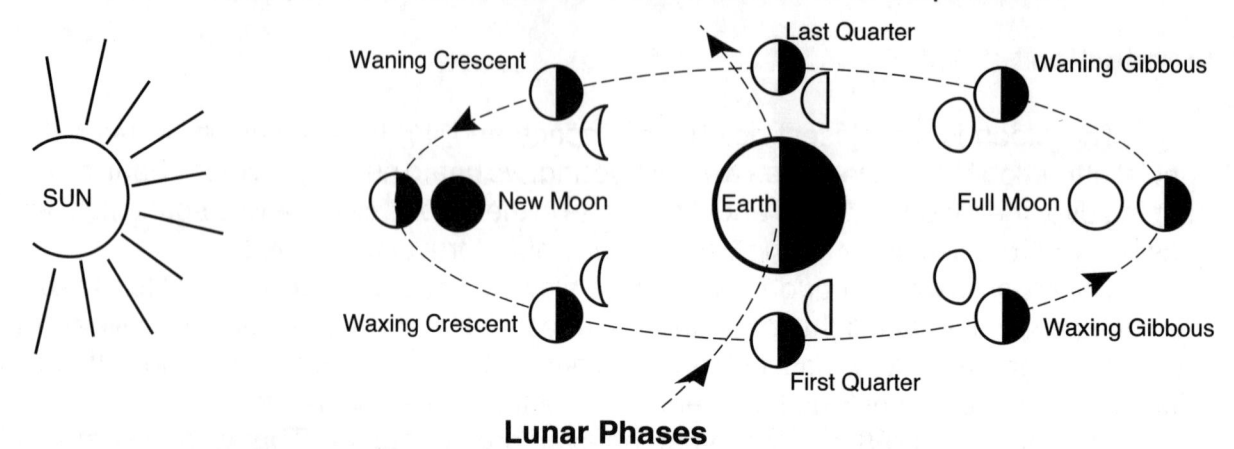

Lunar Phases

The Moon rises and sets at different times when it is in different phases. The table below shows rise and set times for some phases.

PHASE	RISE	SET
New moon	Sunrise	Sunset
First quarter moon	Noon	Midnight
Full moon	Sunset	Sunrise
Last quarter moon	Midnight	Noon

You can sometimes estimate the time of day or night from the position and shape of the Moon if you see it in the sky.

The Moon's Synodic Period

In addition to its sidereal period, the Moon has another period, called the **synodic period.** The synodic period is the time that it takes the Moon to go through a complete cycle of phases. For example, it takes one synodic period to go from a full moon to the next occurrence of a full moon. It takes one synodic period to go from a first quarter moon to the next first quarter moon.

The Moon's synodic period is 29.5 days. To see why it is longer than the sidereal period, look at the diagram below.

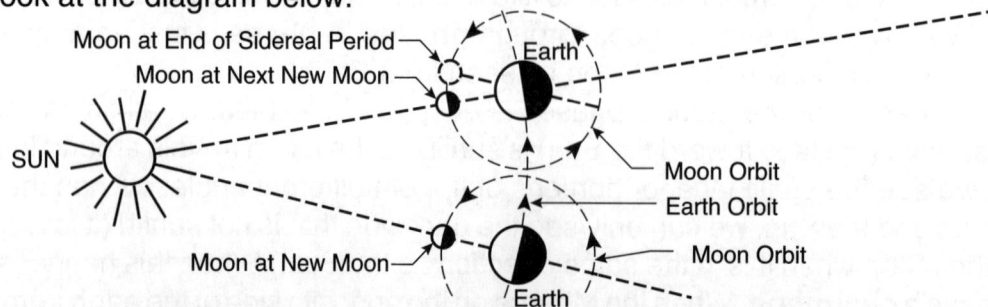

Synodic Period and Sidereal Period of the Moon

Math in Astronomy   Lesson Five: The Moon

The Moon moves completely around in its orbit in 27.3 days. By that time, however, the earth has moved a ways along its orbit. The Sun is no longer in the right position for the Moon to show the same phase that it had 27.3 days before. It has to move a little farther around in its orbit (2.2 more days) to get back to that phase. This is why the synodic period is longer than the sidereal period.

Solar Eclipses

A **solar eclipse** is a blackout of the Sun's light when the Moon passes between the earth and the Sun. The Moon's shadow extends all the way to the earth and causes a brief period of darkness for people who are under it. The Moon's shadow is small, however, and only an area about 273 km wide is in the shadow. As the Earth rotates, the shadow appears to move across the earth. At any one position, a solar eclipse only lasts for a few minutes.

The picture below shows the positions of the Sun, Moon, and Earth during a solar eclipse. **Note that a solar eclipse can only occur during a new moon.**

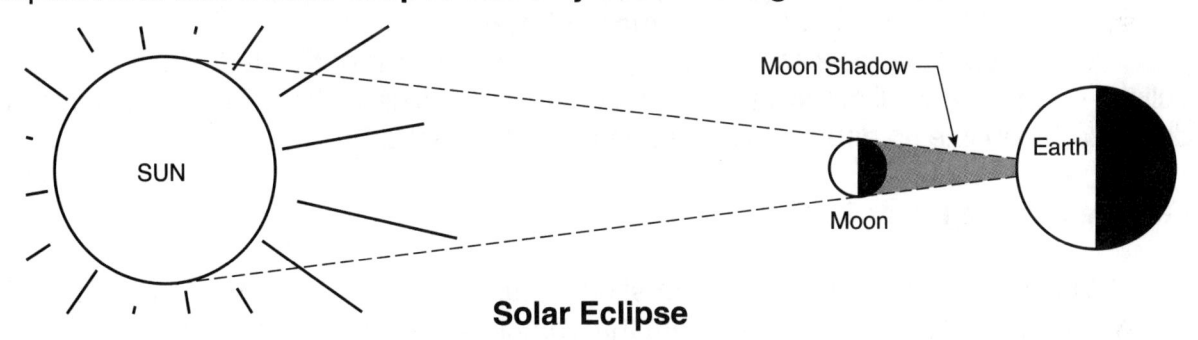

Solar Eclipse

Why do we not have an eclipse every time there is a new moon? We don't because the Moon's orbit is tilted at about 5° with respect to the earth's orbit. Only at certain new moons is the moon aligned just right in its orbit to pass exactly between the earth and the Sun.

The type of solar eclipse that we have just described is a **total solar eclipse.** There are other types. Sometimes the Moon is farther away in its orbit, and it does not completely cover the Sun during the eclipse. A ring of sunlight is seen around the dark shape of the Moon across the Sun. This is called an **annular solar eclipse.**

If you are not in the path of a total solar eclipse or if the Moon and Sun are misaligned so that the Sun is not completely covered, you may see a **partial solar eclipse.** A partial eclipse looks like a "bite" has been taken out of the Sun.

NOTE: You should never look directly at the Sun, even during an eclipse and even with sunglasses on. Serious eye damage and even blindness can result. You may damage your eyes and not even know it until later. Ask your teacher to show you the "projection method" for observing an eclipse.

Lunar Eclipses

When the full moon moves into the shadow of the earth, we experience a **lunar eclipse.** Because the earth's shadow is so large, a lunar eclipse can be seen from the entire dark side of the earth. Also, because the earth's shadow is so large, lunar eclipses can last up to about 100 minutes. If only a portion of the Moon is eclipsed, the eclipse is called a **partial lunar eclipse.**

The picture below shows the alignment of the earth, Sun, and Moon necessary for a lunar eclipse.

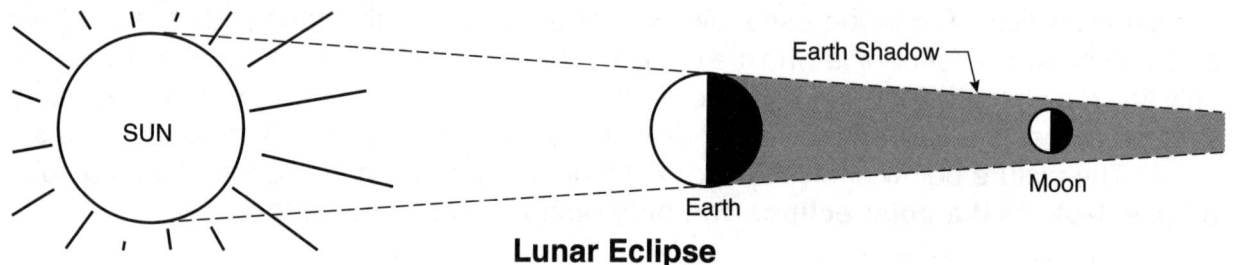

Lunar Eclipse

Obviously, lunar eclipses can only occur during a full moon. We do not have a lunar eclipse during every full moon because the Moon's orbit axis is tilted. Only at certain times does the Moon line up directly in line with the earth and the Sun.

Features on the Lunar Surface

The Moon has a number of interesting features on its surface. Many of them are of interest to us because we can see them through binoculars and small telescopes.

The most obvious features on the lunar surface are the **maria.** These are large, dark areas that were mistaken for seas and oceans by early observers. They named them for bodies of water, using the Latin word for sea, Mare (plural is Maria). We see names like the Mare Tranquilitatis (Sea of Tranquility), the Mare Serenitatis (Sea of Serenity), the Mare Nubium (Sea of Clouds), the Mare Crisium (Sea of Crisis), and others. We know now that these maria are large lava flows that were formed billions of years ago.

The lighter-colored regions of the Moon are known as **highlands.** They are at higher elevations than the maria and are composed of different kinds of rock.

The Moon's surface is covered with **craters.** Craters are roughly circular, bowl-shaped holes. Some of them are more than 160 km wide. They range in depth from a few meters to more than 8,000 meters. It is thought that the craters were formed by the impacts of vast numbers of rocky bodies that crashed into the Moon's surface during its early history. Many of the craters are named after famous scientists.

Key Words

Phase: Appearance or shape of the sunlit portion of the Moon as seen from Earth.

New moon: Phase of the Moon when the Moon is between the earth and the Sun.

Full moon: Phase of the Moon when the Moon is on the opposite side of Earth from the Sun. The entire disk is sunlit.

First quarter moon: Phase of the Moon half way between a new moon and a full moon. It looks like a half circle with the right half sunlit.

Waxing crescent moon: Phase of the Moon between a new moon and a first quarter moon. It looks like a crescent with the right portion sunlit.

Waxing gibbous moon: Phase of the Moon between a first quarter moon and a full moon. More than the right-hand half circle is sunlit.

Last quarter moon: Phase of the Moon half way between a full moon and a new moon. It looks like a half circle with the left half sunlit.

Waning gibbous moon: Phase of the Moon between a full moon and a last quarter moon. More than the left-hand half circle is sunlit.

Waning crescent moon: Phase of the Moon between a last quarter moon and a new moon. It looks like a crescent with the left portion sunlit.

Synodic period: Time required for the Moon to go through a complete cycle of phases.

Solar eclipse: A blackout of the Sun's light when the Moon passes between the earth and the Sun.

Total solar eclipse: A solar eclipse in which the Sun is completely covered by the Moon.

Annular solar eclipse: A solar eclipse when the Moon is farther away in its orbit, and it does not completely cover the Sun. A ring of sunlight is seen around the dark shape of the Moon across the Sun.

Partial solar eclipse: A solar eclipse that is not total. It looks like a "bite" has been taken out of the Sun.

Lunar eclipse: A darkening of the reflection from the Moon when the full moon moves into the earth's shadow.

Partial lunar eclipse: A lunar eclipse when only a portion of the Moon moves into the earth's shadow. It looks like a "bite" has been taken out of the Moon.

Mare: Large, dark plain on the Moon's surface.

Highland: Lighter-colored regions of the Moon at higher elevations than the maria.

Crater: Roughly circular, bowl-shaped holes on the Moon's surface.

Name _____ Date _____

Lesson Five—Questions

1. The appearance or shape of the sunlit portion of the Moon as seen from Earth is called its
   a) perihelion   b) phase   c) temperature   d) phrase

2. The phase of the Moon when the Moon is between the earth and the Sun is called a
   a) new moon   b) blue moon   c) gibbous moon   d) full moon

3. The phase of the moon when the Moon is on the opposite side of the earth from the Sun is called a
   a) new moon   b) blue moon   c) gibbous moon   d) full moon

4. The phase of the Moon halfway between a new moon and the next full moon is called a
   a) new moon   b) first quarter moon
   c) last quarter moon   d) waxing gibbous moon

5. The phase of the Moon half way between a full moon and the next new moon is called a
   a) new moon   b) first quarter moon
   c) last quarter moon   d) waxing gibbous moon

6. The phase of the Moon between a new moon and the next first quarter moon is called a
   a) waxing crescent moon   b) waning crescent moon
   c) waxing gibbous moon   d) waning gibbous moon

7. The phase of the Moon between a first quarter moon and the next full moon is called a
   a) waxing crescent moon   b) waning crescent moon
   c) waxing gibbous moon   d) waning gibbous moon

8. The phase of the Moon between a full moon and the next last quarter moon is called a
   a) waxing crescent moon   b) waning crescent moon
   c) waxing gibbous moon   d) waning gibbous moon

9. The phase of the Moon between a last quarter moon and the next new moon is called a
   a) waxing crescent moon   b) waning crescent moon
   c) waxing gibbous moon   d) waning gibbous moon

10. The time required for the Moon to go through a complete cycle of phases is called the
    a) sidereal period   b) synodic period
    c) rotation period   d) crescent period

Math in Astronomy   Lesson Five—Questions

Name _____ Date _____

Lesson Five—Questions (continued)

11. A blackout of the Sun's light when the Moon passes between the earth and the Sun is called a
    a) lunar eclipse   b) synodic period
    c) gibbous phase   d) solar eclipse

12. A solar eclipse in which the Sun is completely covered by the Moon is called a(n)
    a) lunar eclipse   b) annular solar eclipse
    c) total solar eclipse   d) partial solar eclipse

13. A solar eclipse when the Moon is farther away from the earth in its orbit and does not completely cover the Sun called a(n) _____. (A ring of sunlight is seen around the dark shape of the Moon across the Sun.)
    a) lunar eclipse   b) annular solar eclipse
    c) total solar eclipse   d) partial solar eclipse

14. A solar eclipse that is not total is called a(n) _____. (It looks like a "bite" has been taken out of the Sun.)
    a) lunar eclipse   b) annular solar eclipse
    c) total solar eclipse   d) partial solar eclipse

15. A darkening of the reflection from the Moon when the full moon moves into the earth's shadow is called a(n)
    a) lunar eclipse   b) annular solar eclipse
    c) total solar eclipse   d) partial solar eclipse

16. A lunar eclipse when only a portion of the Moon moves into the earth's shadow is called a(n) _____. (It looks like a "bite" has been taken out of the Moon.)
    a) lunar eclipse   b) annular solar eclipse
    c) total solar eclipse   d) partial lunar eclipse

17. A large, dark plain on the Moon's surface is called a(n)
    a) crater   b) lunar highland
    c) mare   d) eclipse

18. Lighter-colored regions of the Moon at higher elevations than the maria are called
    a) craters   b) maria
    c) icebergs   d) lunar highlands

19. Roughly circular, bowl-shaped holes on the Moon's surface are called
    a) craters   b) maria
    c) icebergs   d) lunar highlands

Lesson Five—Math

Finding Averages

Sometimes, we have several numbers that we want to describe. Instead of writing all of the numbers, we use a single value that we call an **average.** The average of a group of numbers is the sum of the numbers divided by how many numbers there are.

$$\text{average} = \frac{\text{sum}}{\text{number of numbers}}$$

For example, suppose there are five students—Liz, Joe, Sue, Sam, and Meg—in a classroom. Their weights are as follows:

Liz 70 lb
Joe 90 lb
Sue 75 lb
Sam 85 lb
Meg 60 lb

What is the average weight of the students in the classroom?

The **number of weights** (number of numbers) is **5.** The **sum of the weights** (sum of the numbers) is

**sum of weights = 70 lb + 90 lb + 75 lb + 85 lb + 60 lb
= 380 lb**

The **average** is:

$$\text{average weight} = \frac{\text{sum}}{\text{number of numbers}} = \frac{380 \text{ lb}}{5}$$
$$= 76 \text{ lb}$$

The average weight of the students in the classroom is 76 pounds.

The idea of an average can be applied to all sorts of things: an average score for a group of test scores; an average height for a basketball team; or even an average age of the people in some neighborhood or town. Scientists sometimes use the average of a group of measurements in their calculations.

Math in Astronomy   Lesson Five—Math: Finding Averages

Name _____ Date _____

Exercises—Averages

1. Find the average of the following three numbers: 23, 46, and 33. _____

2. Find the average of the following four numbers: 8, 7, 5, and 4. _____

3. Six students received the following scores on a test: 98, 87, 76, 82, 91, and 88. What was the average score on the test? _____

4. The heights of the five starting players on the women's basketball team are: Janet—6.2 ft; Rosa—5.8 ft; Oprah—6.1 ft; Olga—5.9 ft; and Midge—5.5 ft. What is the average height of the five players on the basketball team? _____

5. There are seven girls in a young ladies' literary society. Their ages are 16, 13, 14, 15, 13, 17, and 12. What is the average age of the girls in the literary society? _____

6. Four football players, Jeff, Marcus, Nick, and Carlos, measured the length of a football field. Their measurements are shown below.

    Jeff—301.5 ft   Marcus—298.9 ft
    Nick—299.5 ft   Carlos—302.1 ft

What is the average of their measurements for the football field? _____

7. Four students measured the temperature of a liquid in a science experiment. Their values were: Isaac—308.0 K; Mike—309.1 K; Gina—307.7 K; and Quarette—309.0 K. What was the average of their temperature measurements? _____

© Mark Twain Media, Inc., Publishers

Using Scaling to Find Sizes

In Lesson Two, we used scaling to make a model of the solar system. Scaling has a lot of other uses. In this lesson, we will use it to find the sizes of craters and maria on the Moon.

To see how we can use scaling to find sizes, look at the drawing at the right.

Suppose that this is a family photograph. You know that the father is exactly 1.8 meters tall, and you would like to know the heights of the mother, the kids, and the dog.

You can start by finding a scale for the picture. If you measure the height of the father with a ruler, you will see that it is about 7.2 centimeters. You can get a scale for the picture by dividing the actual height of the father by the measured height in the picture.

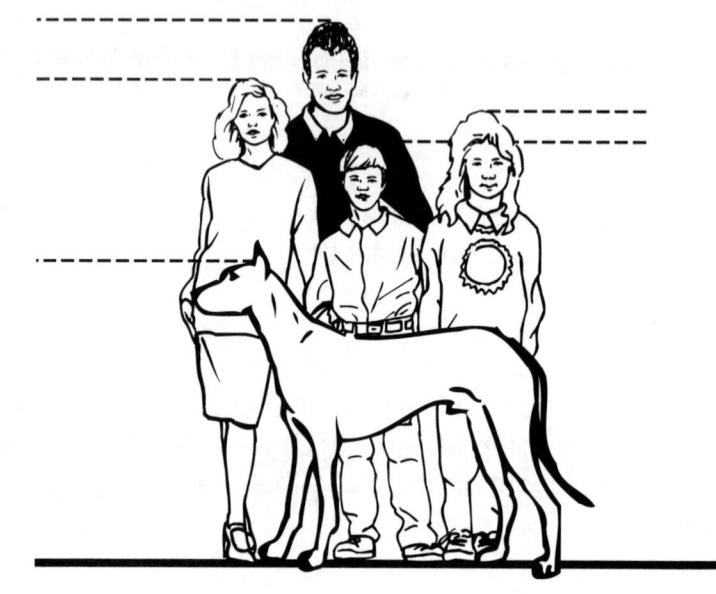

$$\text{scale} = \frac{\text{actual height}}{\text{measured height}}$$
$$= \frac{1.8 \text{ m}}{7.2 \text{ cm}} = 0.25 \text{ m/cm}$$

Then, you can measure the heights of other family members in the picture and multiply the measured heights by the scale to find their actual heights.

actual height = measured height × scale

For example, if you measure the height of the mother in the picture, you will get about 6.4 centimeters. If you multiply her measured height times the scale, you will get her actual height.

mother's height = 6.4 cm × 0.25 m/cm
= 1.6 m

Exercises—Scaling

Test your knowledge of scaling by finding the heights of the boy, the girl, and the dog in the picture.

8. Boy = _____   9. Girl = _____   10. Dog = _____

Math in Astronomy   Lesson Five—Math: Using Scaling to Find Sizes of Craters and Maria on the Moon

Name _____ Date _____

Using Scaling to Find Sizes of Craters and Maria on the Moon

We can use this same scaling process to find the sizes of landforms on the surfaces of moons and planets if we know the diameter of the moon or planet.

First we take a photograph of the moon or planet. Then we get a scale by dividing the actual diameter (in km) by the diameter that we measure in the photograph (in cm).

$$\text{scale} = \frac{\text{actual diameter}}{\text{measured diameter}}$$

Once we have a scale, we can measure the size of some feature in the photograph (in cm), and then multiply it by the scale to find its actual size (in km).

actual size = measured size x scale

The picture on the next page is a sketch made from a photograph of the Moon. Only a few of the many craters and maria that are present on the Moon are shown. The following exercises will ask you to first find a scale and then to find the sizes of some of the craters and maria.

Exercises—Finding Sizes of Maria and Craters

11. The actual diameter of the Moon is about 3,480 kilometers. Measure the diameter of the Moon in the picture in centimeters. Then calculate the scale (in km/cm). _____

12. Measure the diameter of the crater Tycho in the picture (in cm) horizontally. Then measure it again (in cm) vertically. Compute the average of your two measurements. Then multiply the average by your scale factor from Exercise 11 to find the actual (average) diameter of the crater. _____

13. Measure the diameter of the crater Copernicus in the picture (in cm) horizontally. Then measure it again (in cm) vertically. Compute the average of your two measurements. Then multiply the average by your scale factor from Exercise 11 to find the actual (average) diameter of the crater. _____

14. Measure the diameter of the crater Plato in the picture (in cm) horizontally. Then measure it again (in cm) vertically. Compute the average of your two measurements. Then multiply the average by your scale factor from Exercise 11 to find the actual (average) diameter of the crater. _____

15. Measure the distance horizontally across the Mare Crisium in cm. Then measure it again vertically. Then measure it a third time at a 45° angle from lower left to upper right. Finally, measure it a fourth time at a 45° angle from upper left to lower right. Compute the average of your four measurements. Then multiply your average value times your scale from Exercise 11 to find the actual (average) size of the Mare Crisium. _____

© Mark Twain Media, Inc., Publishers

Math in Astronomy   Lesson Five—Math: Using Scaling to Find Sizes of Craters and Maria on the Moon

Name _____ Date _____

Exercises—Finding Sizes of Maria and Craters (cont.)

16. Measure the distance horizontally across the Mare Serenitatis in cm. Then measure it again vertically. Then measure it a third time at a 45° angle from lower left to upper right. Finally, measure it a fourth time at a 45° angle from upper left to lower right. Compute the average of your four measurements. Then multiply your average value times your scale from Exercise 11 to find the actual (average) size of the Mare Serenitatis. _____

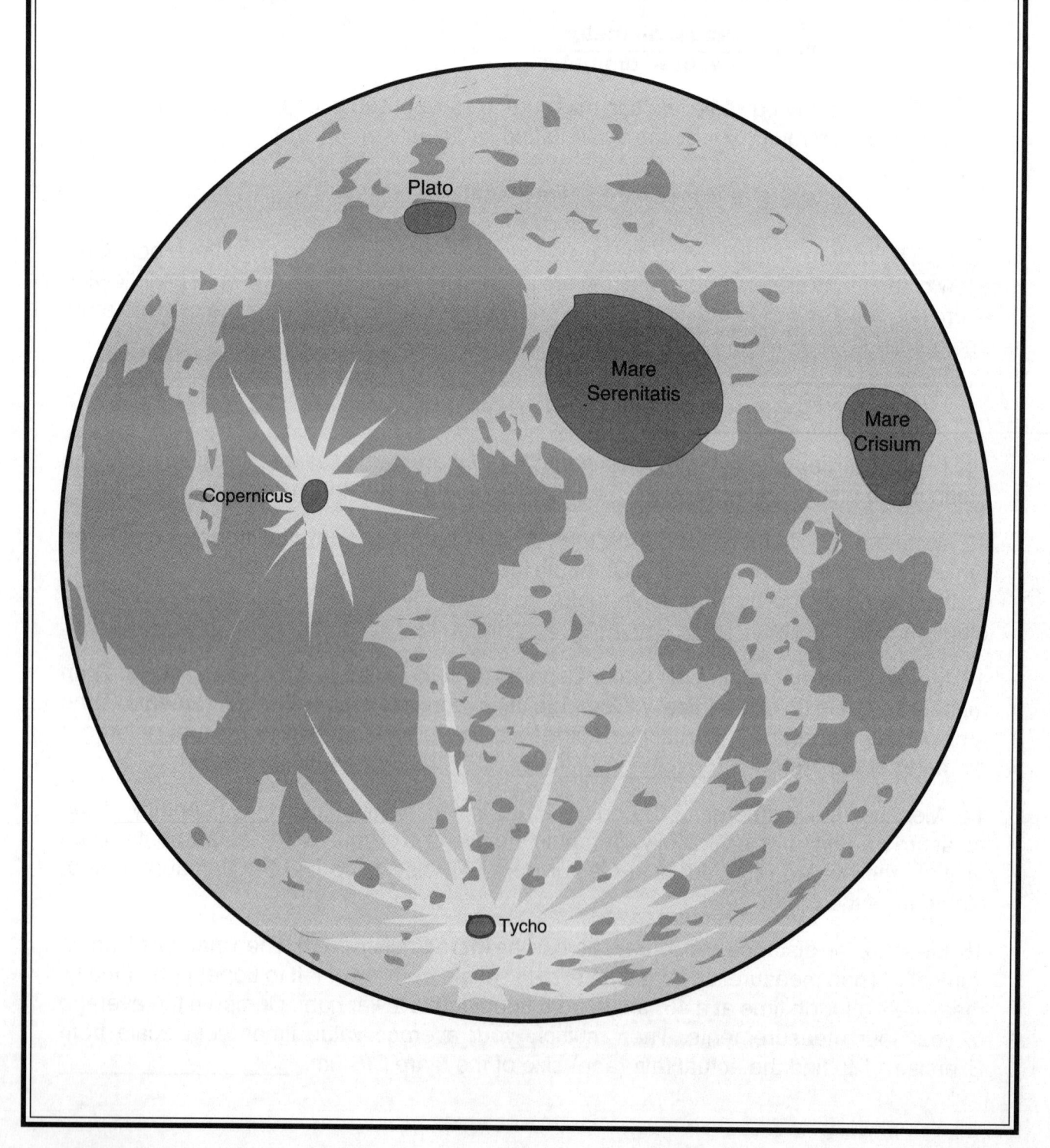

Math in Astronomy
Lesson Five—Teacher's Page

Lesson Five—Teacher's Page

Comment

The picture of the Moon in the Math Section is a rough sketch made from a photograph of the Moon. The numbers that your students get for the sizes of the various lunar features will only be approximate. If you are interested in obtaining more realistic values, you might want to repeat the scaling calculations with an actual photograph of the Moon.

Class Activities

I. The Moon is an excellent object to view with binoculars. A full moon is good when you are looking at the maria, but a first or last quarter moon is usually better if you want to spot craters. The low-angle light casts shadows that make the craters stand out.

II. It might be interesting to find the average weight and the average height of the students in your class.

Answers to Questions (pages 70–71)

1. b   2. a   3. d   4. b   5. c   6. a   7. c   8. d   9. b   10. b

11. d   12. c   13. b   14. d   15. a   16. d   17. c   18. d   19. a

Answers to Mathematical Exercises (pages 72–76)

1. 34   2. 6   3. 87   4. 5.9 ft.   5. 14.3 years
6. 300.5 ft.   7. 308.5 K

8. 1.4 m   9. 1.5 m   10. 1.0 m

11. 211 km/cm

12-16. Teacher check. Measurements and averages will vary slightly.

© Mark Twain Media, Inc., Publishers   77

Lesson Six
Mars

Introduction

Mars is the next planet outward from the Sun after our earth. It is probably the most interesting planet, because in many ways it is similar to Earth. Its rotation period is much the same as Earth's, so its days and nights are almost the same length as ours. Its rotation axis is tilted by about the same amount as Earth's, so it experiences seasons like ours. It even has white regions at its North and South poles that get smaller in the Summer season and in the Winter season.

The most interesting thing about Mars, however, is the belief by some scientists that life may have existed there at some time, and perhaps still does.

Unlike Venus and Mercury, Mars is a superior planet. This means that sometimes it can be seen high in the night sky, long after the Sun has set. When it is near opposition, it is a fun object to observe with small to medium-sized telescopes. Oppositions of Mars occur every two years and 50 days, on the average. The next opposition will be in March of 1997.

Properties of Mars

The properties of Mars are listed in the table below. As you can see, it is small. Its diameter is about half of Earth's and its mass is only a little more than one-tenth of the mass of the Earth. It would take ten Mars's to balance one Earth on a seesaw.

A day on Mars is about 24 hours long. We would feel comfortable getting up at sunrise and going to bed a while after sunset on Mars. A year on Mars would last for 1.88 Earth years, but there would be something like a Summer season and a Winter season. Don't get the idea that Summer would be balmy, however. The average temperature on Mars is only about 253K (-20° C) on the sunlit side and a chilly 133K (-140° C) on the night side.

Mars Properties

Average distance from the Sun	1.52 AU
Orbit period	1.88 years
Orbit eccentricity	0.0934
Inclination of equator to orbit	24°
Sidereal rotation period	24h 37m
Diameter	6,769 km
	0.53 of Earth's
Mass	0.108 Earth masses
Average density	3,940 kg/m^3
Gravity	0.379 of Earth

Relative Sizes of Earth and Mars

Stories about Mars

One reason that we are so fascinated with Mars has to do with some early stories about the planet. In 1877, an Italian astronomer, whose name was **Giovanni Schiaparelli,** thought he saw thin lines on the planet. He called them *canali,* which means channels in Italian, but many people thought he meant **canals.** They thought that Mars might be inhabited by intelligent beings who built canals to carry water from wet parts of Mars to desert regions.

A man named **Percival Lowell** even built an astronomical observatory near Flagstaff, Arizona, to look for life on Mars. Writers began to write stories about adventurous trips to Mars and strange Martian inhabitants. A radio drama was broadcast in 1938 about an invasion of the Earth from Mars. It had people in New Jersey so scared that they packed up their cars and headed for the mountains.

Much of the speculation about life on Mars was put to rest in 1965 when the first American space probe went past Mars. It sent back photographs of a planet on which life of any sort would be difficult. In the past few years, however, scientists have found new evidence for life on Mars.

The Surface of Mars

Mars is sometimes called the Red Planet, and that name is a good one. If you could stand on the surface of Mars, you would see that the soil is reddish in color. The color comes from chemical compounds of iron, similar to the rust that we see here on Earth.

We know this because two American spacecraft, *Viking I* and *Viking II,* landed on the surface of Mars in the Summer of 1976. They photographed the surface of the planet and even analyzed samples of its soil.

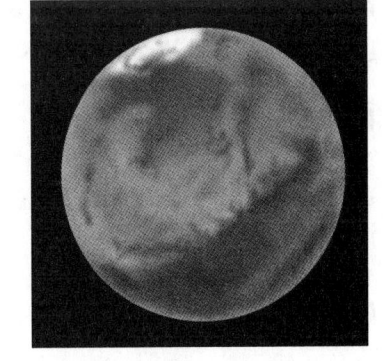

A View of Mars Through a Large Telescope

Other spacecraft have orbited Mars and photographed the planet from many points of view. They have found some amazing things. Mars has craters, canyons, and giant volcanoes. One volcano, called **Olympus Mons,** is 25 kilometers high, higher than Mount Everest here on Earth. A huge canyon, called the **Valles Marineris,** stretches for 4,000 kilometers. It is so big that the Earth's Grand Canyon would be lost in it.

The photographs also showed strange patterns in the land. Some of them looked like they might have come from big floods, but others looked like the patterns of streams and rivers that we see here on Earth. Most scientists agree that these patterns mean that Mars must have had rain and flowing streams of water at some time in its history.

Finally, there are **polar caps** on Mars. They are white areas near the North and South Poles of Mars, and for a while observers thought they might be snow and ice like the ones on Earth. After more measurements, they realized that they were carbon dioxide ice (what we call "dry ice" here on Earth) with a smaller water-ice part underneath. The polar cap at the North Pole gets larger when it is Winter in the Northern Hemisphere and smaller when it is Summer there. The one near the South Pole gets larger when it is Winter in the Southern Hemisphere and smaller when it is Summer there.

The Atmosphere of Mars

Like the Earth and Venus, Mars has an atmosphere. It pushes down on Mars' surface with only about one hundredth as much pressure as our atmosphere does here on Earth. It is composed primarily of carbon dioxide.

The air on Mars is thin, but winds can still blow, and dust storms can whip the reddish dust about. Sometimes a dust storm can envelope the whole planet.

The Moons of Mars

Mars has two moons, **Phobos** and **Deimos,** that orbit around the planet. Both of them are very small compared to our moon. They are irregularly shaped (not spherical like the planets and our moon) and only a few kilometers across. Phobos is slightly larger than Deimos, and its orbit is much closer to Mars.

Deimos orbits Mars once every 30h 18m. **Phobos,** however, goes all the way around in its orbit in a little less than eight hours. Its orbit speed is so fast that it outruns the rotation of Mars. If you were on the surface of Mars, you would see Phobos rise in the West, go the "wrong way" across the sky, and set in the East.

Key Terms

Giovanni Schiaparelli: An Italian astronomer who thought he observed channels on Mars in 1877.

Canals: Long, thin features that people thought they saw on the surface of Mars.

Percival Lowell: Astronomer who built an observatory near Flagstaff, Arizona, to search for evidence of life on Mars.

Olympus Mons: Giant volcano on Mars. The highest mountain in the solar system.

Valles Marineris: 4,000-km-long canyon on the surface of Mars.

Polar caps: White areas near the North and South Poles of Mars.

Phobos: The larger of Mars' two moons. Its orbit is closest to Mars.

Deimos: The smaller of Mars' two moons. Its orbit is farthest from Mars.

Math in Astronomy   Lesson Six—Questions

Name _____ Date _____

Lesson Six—Questions

1. The Italian astronomer who thought he observed channels on Mars in 1877 was

    a) Percival Lowell   b) Galileo   c) Giovanni Schiaparelli   d) Carl Sagan

2. Long, thin features that people thought they saw on the surface of Mars are called

    a) pipelines   b) cracks   c) latitudes   d) canals

3. The astronomer who built an observatory near Flagstaff, Arizona, to search for evidence of life on Mars was

    a) Percival Lowell   b) Galileo   c) Giovanni Schiaparelli   d) Carl Sagan

4. The giant volcano on Mars is called _____ . It is the highest mountain in the solar system.
    a) Olympus Mons   b) Maxwell Montes c) Mount Olympus   d) Valles Marineris

5. The 4,000-km-long canyon on the surface of Mars is called

    a) Olympus Mons   b) Grand Canyon   c) Royal Gorge   d) Valles Marineris

6. The white areas near the North and South poles of Mars are called the

    a) Valles Marineris b) polar caps   c) white ovals   d) Olympus Mons

7. The larger of Mars' two moons is called _____ . Its orbit is closest to Mars.

    a) Deimos   b) Phobos   c) Luna   d) Olympus Mons

8. The smaller of Mars' two moons is called _____ . Its orbit is furthest from Mars.

    a) Deimos   b) Phobos   c) Luna   d) Olympus Mons

9. The average distance of Mars from the Sun is 1.52 AU, and the eccentricity of its orbit is 0.0934.
    a) What is the perihelion distance of Mars from the Sun? _____
    b) What is the aphelion distance of Mars from the Sun? _____

10. Suppose the surface temperature on the sunlit side of Mars is 250 K.
    a) What would that temperature be in degrees Celsius? _____
    b) What would that temperature be in degrees Fahrenheit? _____

© Mark Twain Media, Inc., Publishers

Lesson Six—Math

Travel Times to the Planets

Much of what we know about the planets has been gained by sending unmanned space probes to them. We have sent spacecraft to all of the planets except Pluto. In the future we may launch manned spaceships to Mars.

The trips to the moon made by the astronauts from 1969 to 1972 took only three days. Travel times to the planets are much longer, particularly if a path that uses the least amount of energy is taken.

The least energy path to a planet involves launching a spacecraft from Earth and then putting it into an elliptical orbit that touches Earth's orbit on one end and touches the orbit of the other planet on the other end. Such an orbit (for a trip to an inferior planet) is shown below.

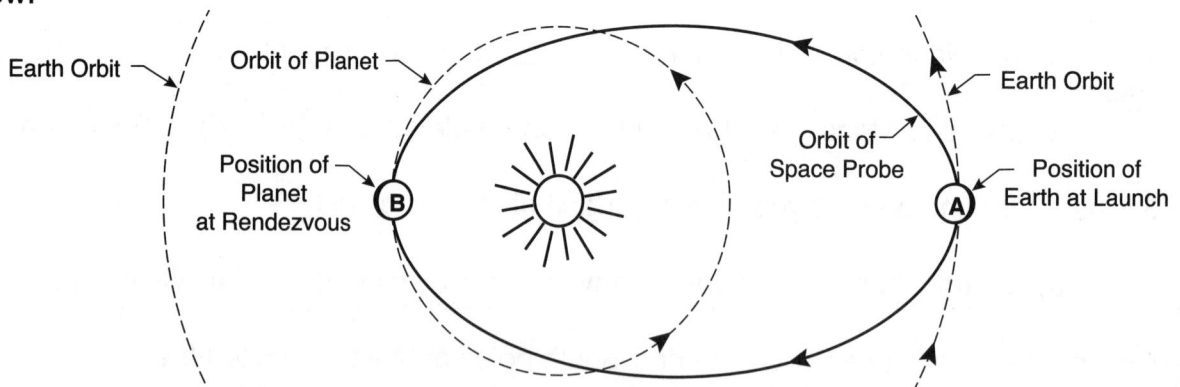

To send a spaceship to an inferior planet along the orbit shown above would take careful timing. We would have to launch the spacecraft when the earth was at point A in its orbit. We would first place it in an orbit around the earth. Then we would adjust its speed so that it would go into the long elliptical orbit that would carry it from point A to a rendezvous point with the inferior planet at point B. During the trip from A to B, no rocket power would be necessary. The spaceship would coast in its orbit.

If we timed our mission right, the inferior planet would be at point B when our spacecraft got there. We could then adjust the speed of our spacecraft again so that the craft would go into orbit around the planet.

We can use Kepler's Third Law to find how long it would take to go from the earth to an inferior planet along such an orbit. The semimajor axis of the orbit is one-half of the sum of the semimajor axis of the earth's orbit (1 AU) and the semimajor axis of the orbit of the planet.

$$\text{semimajor axis} = \frac{1 + \text{semimajor axis of planet}}{2}$$

Once we knew the semimajor axis, we could calculate the period from Kepler's Third Law.

$$\text{period} = \sqrt{(\text{semimajor axis})^3}$$

Math in Astronomy   Lesson Six—Math: Travel Times to the Planets

Name _____   Date _____

The period is the time that it would take to travel all of the way around the elliptical orbit. To get to the planet, we would only need to travel half that amount of time.

travel time = $\dfrac{\text{period}}{2}$

Exercises—Travel Times to Inferior Planets

1. Find the travel time from the earth to Mercury along an orbit like the one shown on the previous page. Mercury's orbit has a semimajor axis of 0.38 AU. Sketch the orbits of the earth and Mercury, and the orbit that would be followed by the spaceship. _____

2. Find the travel time from the earth to Venus along an orbit like the one shown on the previous page. Venus's orbit has a semimajor axis of 0.72 AU. Sketch the orbits of the earth and Venus and the orbit that would be followed by the spaceship. _____

The same calculation that we used for trips to inferior planets also applies to trips to superior planets.

Exercises—Travel Times to Superior Planets

3. Sketch the orbits of the earth and Mars and the elliptical orbit that would be followed by a spaceship going from Earth to Mars. (Remember, it must just touch the orbits of Earth and Mars at its opposite ends.) Then find the travel time from Earth to Mars along the orbit. Mars' orbit has a semimajor axis of 1.52 AU. _____

4. Sketch the orbits of the Earth and Jupiter and the elliptical orbit that would be followed by a spaceship going from Earth to Jupiter. Then find the travel time from the Earth to Jupiter along the orbit. Jupiter's orbit has a semimajor axis of 5.20 AU. _____

Name _____ Date _____

Communication Times to the Planets

The planets are far enough away that we have to take the speed of light into account when we send radio signals to them or receive signals from spacecraft that are in orbit around them or on their surfaces. Radio signals travel at the speed of light, which is about **0.0020 AU per second.** To find the time (in seconds) that a radio signal would take to travel a certain distance, simply divide that distance (in AU) by the speed of light (in AU/second).

$$\text{time} = \frac{\text{distance}}{0.0020 \text{ AU/sec}}$$

For example, suppose that a spaceship was 6 AU from Earth. How long would it take a radio signal to go from Earth to the spaceship?

$$\text{time} = \frac{6 \text{ AU}}{0.0020 \text{ AU/sec}}$$

$$= 3{,}000 \text{ sec}$$

$$= \frac{3{,}000}{60} \text{ min}$$

$$= 50 \text{ min}$$

Exercises—Communication Times

5. How long would it take a radio signal to travel from the earth to the Sun? _____

6. How long would it take a radio signal to travel from Earth to Mercury when the earth and Mercury were in the positions shown below? _____

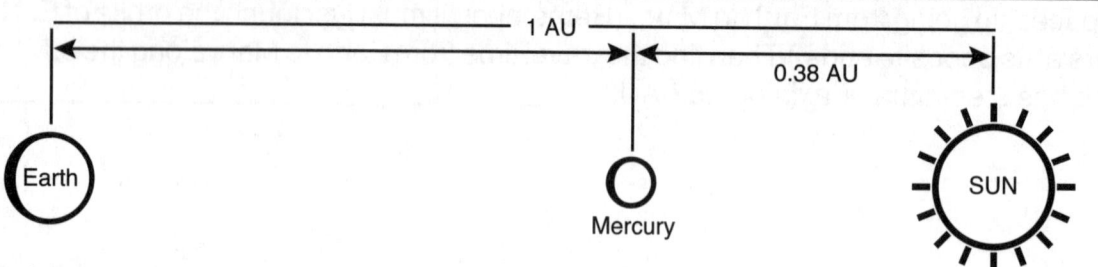

Earth ←——— 1 AU ———→ Mercury ←— 0.38 AU —→ SUN

7. How long would it take a radio signal to travel from Venus to Earth when the earth and Venus were in the positions shown below? _____

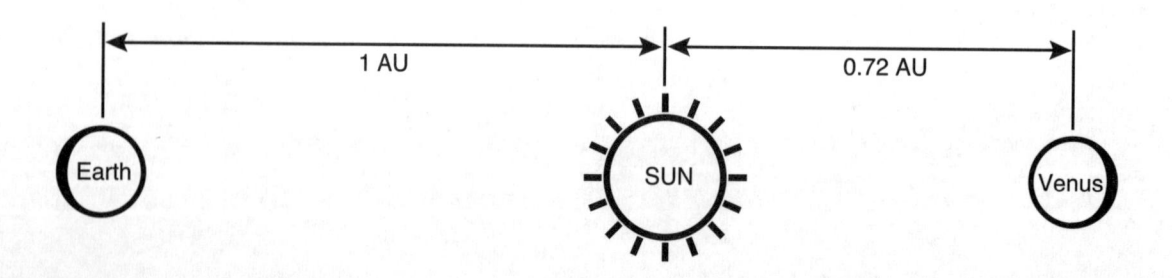

Earth ←——— 1 AU ———→ SUN ←— 0.72 AU —→ Venus

Name _____ Date _____

Exercises—Communication Times (continued)

8. How long would it take a radio signal to travel from Venus to Earth when the earth and Venus were in the positions shown below? _____

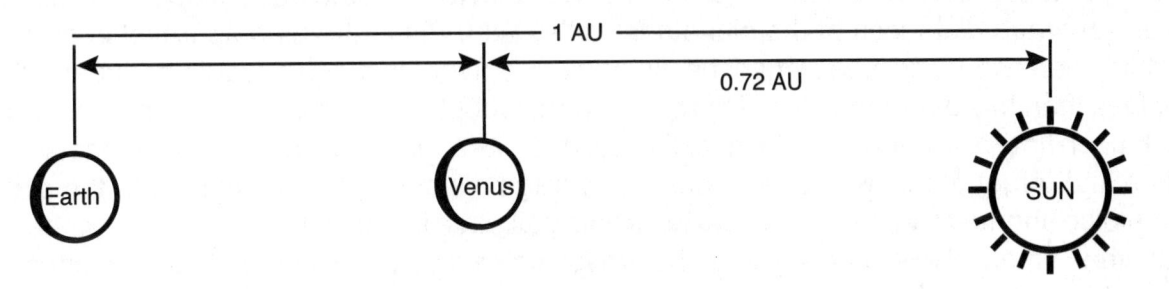

```
|←———————————— 1 AU ————————————→|
                                 |←——— 0.72 AU ———→|
  Earth              Venus                           SUN
```

9. How long would it take a radio signal to travel from Mars to Earth when Mars and the earth were in the positions shown below? _____

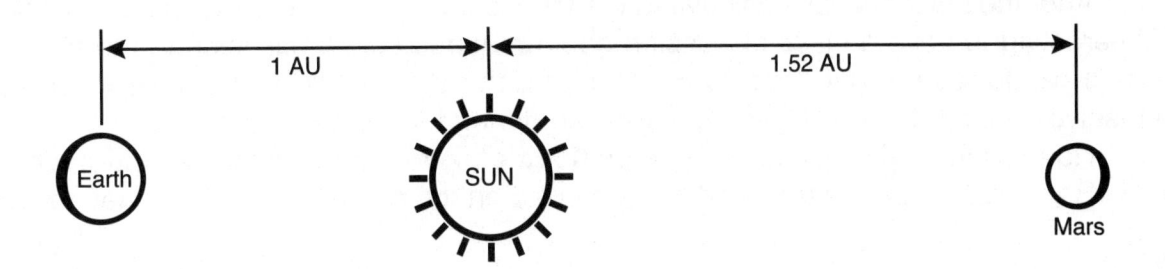

```
|←——— 1 AU ———→|←——————— 1.52 AU ———————→|
  Earth              SUN                          Mars
```

10. How long would it take a radio signal to travel from Earth to Mars when Earth and Mars were in the positions shown below? _____

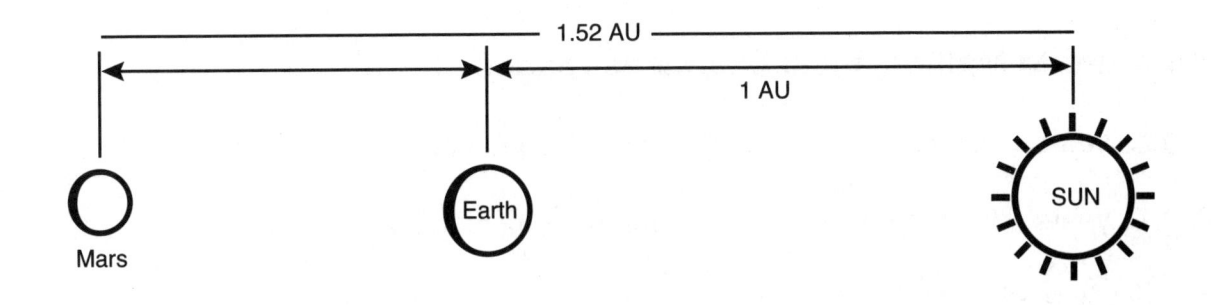

```
|←——————————— 1.52 AU ———————————→|
                |←———— 1 AU ————→|
  Mars              Earth                         SUN
```

Math in Astronomy Lesson Six—Teacher's Page

Lesson Six—Teacher's Page

Comment

The distances in the math section that we used to calculate travel times to the planets were average distances of the planets from the Sun. Since the planets travel in elliptical orbits, there would be times when the distance of a planet to the Sun would be greater than or less than this average value. Therefore, there would be favorable configurations when, at launch time, the earth was closer to the Sun than its average distance and, at rendezvous time, the target planet was closer than its average distance. The travel time to the planet would be shorter than the values that we would calculate in the math section. Unfortunately, at other times, the distances and the travel times could be larger than the ones we calculated.

Class Activity

After the students have calculated the time to travel from the Earth to Mars it would be interesting to have a class discussion about manned space missions to the planets. Since it would take a long time to get to Mars, the astronauts would have to occupy themselves during the trip. What might they do during this time period?

How much food would be necessary (for the trip to Mars and for the return trip)?

How much water would need to be carried on the spacecraft? Could they recycle water?

Answer to Questions (page 81)

1. c   2. d   3. a   4. a   5. d   6. b   7. b   8. a   9a. 1.38 AU   9b. 1.66 AU
10a. -23° C   10 b. -9.4°F

Answers to Mathematical Exercises (pages 82–85)

1. 0.28 years = 3.4 months

2. 0.40 years = 4.8 months

3. 0.71 years = 8.5 months

4. 2.7 years = ~32 months

5. 500 sec = ~8.3 min
6. 310 sec = ~5.2 min
7. 860 sec = ~14 min
8. 140 sec = ~2.3 min
9. 1,260 sec = 21 min
10. 260 sec = ~4.3 min

© Mark Twain Media, Inc., Publishers

Lesson Seven: The Outer Planets

The four planets nearest the Sun, Mercury, Venus, Earth, and Mars are alike in many ways. They are close to the Sun, they have solid surfaces, and, compared to the planets that we will investigate in this lesson, they are small. They are sometimes called the **inner planets** because they are close to the Sun or the **terrestrial planets** because they are like our Earth in some ways. (*Terrestrial* means "earth-like" to astronomers.)

The next four planets outward from the Sun, Jupiter, Saturn, Uranus, and Neptune, are much different. They are very large and have thick atmospheres that are made up of gases that are unlike the ones found on the inner planets. They do not have surfaces like the terrestrial planets have. We sometimes call them the **gas giant planets** or the **outer planets.**

Most of what we know about the gas giant planets has come from the *Voyager 1* and *Voyager 2* space probes that we sent there from 1979 to 1986. These spacecraft carried cameras and sensitive measuring instruments that gave us lots of information.

Pluto, the farthest planet from the Sun, is much different from any of the other planets in the solar system. It has a surface, but it is so cold that its atmosphere is frozen on the ground. Its small size and its strange orbit have led some astronomers to believe that it is not really a planet.

Jupiter

The giant in the solar system is the planet Jupiter. Its diameter is eleven times the diameter of Earth, and it contains more mass than all of the other planets in the solar system put together. It would take 318 Earths to balance one Jupiter on a seesaw.

Relative Sizes of Earth and Jupiter

Jupiter spins rapidly about its axis. A day on Jupiter is only about 9 hours and 50 minutes long. The period of daylight is half of this, so daytime on Jupiter only lasts for a little less than five hours. You would have to rush to get much done during a Jupiter day. Jupiter spins so fast that it bulges outward at its equator.

Jupiter has short days but it has a very long year. It takes 11.9 Earth years for Jupiter to go around the Sun once. This means that a Jupiter year is 11.9 times as long as an Earth year. If you were 35 years old on Earth, you would only be a little less than three Jupiter years old.

Jupiter Properties

Average distance from the Sun	5.20 AU
Orbit period	11.9 years
Orbit eccentricity	0.0484
Sidereal rotation period	9h 50.5m
Diameter	11.2 Earth diameters
Mass	318 Earth masses
Average density	1,340 kg/m^3

Jupiter is best observed when it is at opposition. Through a telescope, it shows a cloudy disk with horizontal **belts** and **zones.** The **belts** are dark cloud bands across the disk of the planet, and the **zones** are lighter bands. Views from powerful telescopes and spacecraft have shown red, brown, blue, yellow, and white colors in the belts and zones.

The belts and zones result from motions of clouds in Jupiter's atmosphere. The atmosphere is primarily hydrogen and helium, and the clouds are made up of ammonia, water, and other chemicals. Methane (the stuff that we call natural gas) is also present in Jupiter's atmosphere. Jupiter is cold—about 150 K at its cloud tops. It warms up, however, as you descend deeper into the clouds.

One of the interesting features in Jupiter's clouds is called the **Great Red Spot.** It is a giant, swirling high pressure storm in the atmosphere of Jupiter. Although it sometimes increases or decreases in intensity, it has been observed for more than 300 years. At its largest size, it could contain several earths. It appears red in color when observed through high power telescopes or the *Voyagers'* cameras.

When we view Jupiter, we are only seeing the outermost layer of clouds in its atmosphere. If we could go deeper in the atmosphere, we would feel the gas getting denser and denser, until it would seem like a liquid. The liquid, however, would have no surface.

View of Jupiter

The Moons of Jupiter

Jupiter has a total of 16 moons. All but four of them are very small and not easily observable from Earth. Some are so tiny that they have only been seen by space probes. Four of the moons, however, are large enough to be seen through even small telescopes. They were visible to **Galileo Galilei** back in 1610 when he built what was probably the first astronomical telescope and observed Jupiter. They are called the **Galilean moons,** and they look like tiny dots near the big disk of Jupiter.

Jupiter's Galilean moons are called **Io, Europa, Ganymede,** and **Callisto.** All of them except Europa are larger than our Moon, but all of them except Io are less dense than our Moon.

Io is the closest to Jupiter and is a very strange moon. Its surface is covered with small volcanoes that spew out sulfur and sulfur compounds. Various-colored deposits of sulfur coat its surface.

Europa is the next big moon outward from Jupiter. It is a rocky moon covered by a thin layer of water ice. Lines that are seen on the surface are cracks in the ice. There may be a layer of water or slush beneath the ice.

Ganymede is the next large moon outward from Jupiter. Like Europa, it is a rocky world, but with a thick layer of ice on its surface. Ganymede is the largest moon in the solar system.

Callisto is the outermost of Jupiter's big, Galilean moons. Like Ganymede, it has a rocky core topped by a thick layer of ice. Its surface is scarred by craters from the impacts of long-ago collisions with meteorites.

Saturn

Perhaps the most interesting planet to view through a telescope is Saturn. Galileo Galilei observed it through his primitive astronomical telescope in 1610. It is the second largest planet in the solar system.

The reason that Saturn is so interesting is that it is surrounded by rings. They are bright and wheel-like. From Earth, three rings are visible, the outermost **A ring,** the intermediate **B ring,** and the innermost **C ring.** A dark ring, called the **Cassinni division,** separates the A ring from the B ring. More rings are visible in photographs taken by the *Voyager* spacecraft. The collection of rings is more than 200,000 km in diameter but only hundreds of meters thick. Some of the rings are made up of ice balls and ice-covered chunks of rock. Others are composed of smaller, dust-like particles. (Jupiter also has a faint ring around it, but it is not visible from Earth.)

View of Saturn

Saturn revolves around the Sun in an orbit that is more than twice as big as Jupiter's orbit. It rotates on its axis in a little over ten and one-half hours, a little longer than Jupiter takes. Like Jupiter, however, it rotates fast enough to cause a bulge at its equator. Saturn's year is more than 29 Earth years long, almost three times as long as Jupiter's year.

Saturn is the least dense planet in the solar system. Its density is only 710 kg/m^3, less than that of water. If you had a giant bathtub full of water and you dropped Saturn into it, Saturn would float.

Saturn Properties

Average distance from the Sun	9.54 AU
Orbit period	29.46 years
Orbit eccentricity	0.0560
Sidereal rotation period	10h 39.25m
Diameter	9.45 Earth diameters
Mass	95.2 Earth masses
Average density	710 kg/m^3

Like Jupiter, Saturn has an atmosphere that is mostly hydrogen and helium, and it has clouds that are composed of ammonia, methane, water, and other chemicals. Also, like Jupiter, the density of the atmosphere increases as you go deeper into it. Eventually, it becomes liquid. Saturn's temperature is about 90 K at its cloud tops.

Saturn's Big Moon

Saturn has a total of 20 moons. The most interesting one, however, is **Titan.** It is much larger than any of Saturn's other moons, and it is the only moon in the solar system to have much of an atmosphere. Its atmosphere is mostly nitrogen (the same gas that makes up most of Earth's atmosphere.) It contains other chemicals, however, such as propane (the gas in rural gas tanks) and carbon monoxide.

Uranus and Neptune

Uranus and Neptune are the other two gas giant planets in the solar system. Uranus was discovered in 1781 by an English astronomer, **William Herschel.** It is the third largest planet in the solar system, with a diameter that is about four (actually 4.01) Earth diameters. Its mass is about 14.5 Earth masses. Like the other gas giants, its atmosphere is hydrogen and helium with small amounts of methane and maybe ammonia. It is an icy 50 K at the top of its atmosphere.

Uranus is about 19.2 AU from the Sun and takes about 84 Earth years to complete one revolution in its orbit. Its rotation period is 17h 14m. Its rotation axis, however, is tilted at almost 98° so that it lays on its side in its orbit.

Neptune is a little smaller but a little heavier than Uranus (its diameter is about 3.9 Earth diameters, but its mass is 17.2 times the mass of the earth). Its orbit is at 30 AU from the Sun and it takes almost 165 years to go around its orbit once. It rotates once in 16h 3m. Its atmosphere is composed of hydrogen and helium with some methane. It is about 60 K at the top of its atmosphere.

The existence of Neptune was predicted mathematically. Observations indicated that there were irregularities in Uranus' orbit that could be caused by the gravity of another planet. Calculations showed that a planet with the size and orbit of Neptune could cause these irregularities. Astronomers found the planet Neptune at the position predicted by the mathematics.

Uranus has 15 moons, but Neptune has only eight. Neptune's biggest moon, Triton, has a small amount of atmosphere, about 1/100,000 of the amount that Earth has.

Both Uranus and Neptune have faint rings that are not visible from Earth.

Pluto

Pluto is the lonely planet on the outskirts of the solar system. It is a place of unbelievable cold. It is so small (its diameter is 0.19 Earth diameters, and its mass is 0.0025 Earth masses) and so far away (about 39 AU from the Sun) that it was not discovered until 1930. It takes an incredible 248 years to circle the Sun. We think that its temperature is about 40 K and that its atmosphere may lay frozen on its surface.

Pluto's orbit is very eccentric (0.248) so that, at times, it is actually closer to the Sun than Neptune. Pluto will be closer to the Sun than Neptune until 1999.

Pluto has a single moon, called **Charon** (pronounced Karen), that is almost as big as the planet. Some astronomers prefer to think of Pluto as a "double asteroid" (see Lesson Eight) instead of a planet and a moon.

Math in Astronomy   Lesson Seven: The Outer Planets

Key Terms

Terrestrial planets: The four planets, Mercury, Venus, Earth, and Mars, that are closest to the Sun. They all have earth-like properties.

Inner planets: The terrestrial planets.

Gas giant planets: The four planets, Jupiter, Saturn, Uranus, and Neptune, that are located at large distances from the Sun. They are large and have gaseous atmospheres.

Outer planets: The gas giant planets.

Belts: Dark-colored cloud bands in the atmosphere of Jupiter.

Zones: Light-colored cloud bands in the atmosphere of Jupiter.

Galileo Galilei: Italian scientist and philosopher who lived from 1564 to 1642. He built the first astronomical telescope in 1610 and observed the planets, including Jupiter and Saturn. He is known by his first name, Galileo.

Galilean moons: The four big moons of Jupiter that were first observed by Galileo.

Io: The innermost Galilean moon of Jupiter. Its surface is covered with small volcanoes that spew out sulfur and sulfur compounds.

Europa: The next innermost Galilean moon of Jupiter. It is a rocky moon covered by a thin layer of water ice. Lines that are seen on the surface are cracks in the ice.

Ganymede: The next to outermost Galilean moon of Jupiter and the largest moon in the solar system. It is a rocky moon, with a thick layer of ice on its surface.

Callisto: The outermost Galilean moon of Jupiter. It has a rocky core covered by a thick layer of ice that is scarred by many craters.

A ring: The outermost bright ring of Saturn that is visible from Earth.

B ring: The bright ring of Saturn between the A ring and the C ring. It is visible from Earth.

C ring: The innermost bright ring of Saturn that is visible from Earth.

Cassinni division: Dark ring of Saturn between the A ring and the B ring.

Titan: Saturn's largest moon and the only moon in the solar system to have an extensive atmosphere.

William Herschel: English astronomer who lived from 1738 to 1822. He discovered Uranus in 1781.

Charon: Moon of the planet Pluto.

Math in Astronomy   Lesson Seven—Questions

Name _____ Date _____

Lesson Seven—Questions

1. The four planets, Mercury, Venus, Earth, and Mars that are closest to the Sun are called the _____ . They all have earth-like properties.

    a) gas giant planets   b) terrestrial planets
    c) outer planets   d) inferior planets

2. The terrestrial planets are also called the

    a) inner planets   b) inferior planets
    c) blue planets   d) gas giant planets

3. The four planets, Jupiter, Saturn, Uranus, and Neptune, that are located at large distances from the Sun are called the _____ . They are large and have thick gaseous atmospheres.

    a) inner planets   b) terrestrial planets
    b) inferior planets   d) gas giant planets

4. The gas giant planets are also called the

    a) inner planets   b) inferior planets
    c) outer planets   d) terrestrial planets

5. The dark-colored cloud bands in the atmosphere of Jupiter are called

    a) red spots   b) belts   c) zones   d) polar caps

6. The light-colored cloud bands in the atmosphere of Jupiter are called

    a) red spots   b) belts   c) zones   d) polar caps

7. The Italian scientist and philosopher who lived from 1564 to 1642 was _____ . He built the first astronomical telescope in 1610 and observed the planets, including Jupiter and Saturn.

    a) Giovanni Schiaperelli   b) William Herschel
    c) Albert Einstein   d) Galileo

8. The innermost Galilean moon of Jupiter is called

    a) Europa   b) Io   c) Callisto   d) Ganymede

Name _____ Date _____

Lesson Seven—Questions (continued)

9. The next to innermost Galilean moon of Jupiter is called

    a) Europa   b) Io   c) Callisto   d) Ganymede

10. The next to outermost Galilean moon of Jupiter and the largest moon in the solar system is called

    a) Europa   b) Io   c) Callisto   d) Ganymede

11. The outermost Galilean moon of Jupiter is called

    a) Europa   b) Io   c) Callisto   d) Ganymede

12. The outermost bright ring of Saturn that is visible from Earth is the

    a) Cassinni division   b) B ring   c) C ring   d) A ring

13. The bright ring of Saturn between the A ring and the C ring is the _____. It is visible from Earth.

    a) Cassinni division   b) B ring   c) C ring   d) A ring

14. The innermost bright ring of Saturn that is visible from Earth is the

    a) Cassinni division   b) B ring   c) C ring   d) A ring

15. The dark ring of Saturn between the A ring and the B ring is the

    a) Cassinni division   b) B ring   c) C ring   d) Aring

16. Saturn's largest moon and the only moon in the solar system to have an extensive atmosphere is called

    a) Phobos   b) Ganymede   c) Io   d) Titan

17. The English astronomer who discovered Uranus in 1781 was

    a) Giovanni Schiaperelli   b) William Herschel   c) Galileo   d) Carl Sagan

18. The moon of the planet Pluto is

    a) Titan   b) Io   c) Charon   d) Phobos

Lesson Seven—Math

The Mass of Jupiter and Other Planets

A modified form of Kepler's Third Law, derived by the English physicist Isaac Newton (1642–1727), allows us to calculate the mass of a planet if we know the size and period of the orbit of one of its moons. The modified form of Kepler's law is shown below.

$$\text{mass of planet} + \text{mass of moon} = \frac{(\text{semimajor axis of moon's orbit})^3}{(\text{period of moon})^2}$$

The **mass of the planet** and the **mass of the moon** are expressed as numbers of Earth masses.

The **semimajor axis of the moon's orbit** is expressed as some number of Earth-Moon distances. (Earth-Moon distance = 3.84×10^5 km.)

The **period of the moon** is expressed as some number of Earth Moon periods (period of Earth's Moon = 27.3 days.)

If the mass of the planet's moon is really small compared to the mass of the planet, the Modified Third Law can be written in the approximate form below.

$$\text{mass of planet} = \frac{(\text{semimajor axis of moon's orbit})^3}{(\text{period of moon})^2}$$

We can measure the period and semimajor axis of a moon's orbit, convert them to the units that we need, and then easily compute the mass of a planet in Earth masses. We will use the approximate form of the Modified Third Law to show how the masses of some of the planets can be calculated.

EXAMPLE: The moon Io orbits Jupiter in an orbit whose semimajor axis is 1.10 Earth-Moon distances, and whose orbit period is 0.0647 Earth Moon periods. Find the mass of Jupiter.

$$\text{mass} = \frac{1.10^3}{0.0647^2}$$

$$= 317.9$$

$$\sim 318 \text{ Earth masses}$$

Exercises—Finding the Masses of Planets

1. The moon Ariel orbits Uranus in an orbit whose semimajor axis is 0.497 Earth-Moon distances, and whose orbit period is 0.0922 Earth Moon periods. Find the mass of Uranus in Earth masses. _____

2. The moon Nereid orbits Neptune in an orbit whose semimajor axis is 14.4 Earth-Moon distances, and whose orbit period is 13.2 Earth Moon periods. Find the mass of Neptune in Earth masses. _____

3. The moon Iapetus orbits Saturn in an orbit whose semimajor axis is 9.26 Earth-Moon distances, and whose orbit period is 2.89 Earth Moon periods. Find the mass of Saturn in Earth masses. _____

4. The moon Deimos orbits Mars in an orbit whose semimajor axis is 0.0610 Earth-Moon distances, and whose orbit period is 0.0462 Earth Moon periods. Find the mass of Mars in Earth masses. _____

Math in Astronomy

Lesson Seven—Teacher's Page

Comment

The outer planets are all superior planets. Therefore, they can appear in good viewing positions late at night. Only Jupiter and Saturn, however, can be seen easily with the naked eye. Uranus can be seen, but only with difficulty, and you have to know exactly where to look.

When they are near opposition, Jupiter and Saturn are very bright sky objects. Students should have no trouble finding them at night if given a region of the sky to look toward. These two planets are the most rewarding ones to view through small telescopes. Even a very modest telescope will show Jupiter as a disk and its four Galilean moons. A slightly larger one will reveal belts and zones in Jupiter's clouds and the rings of Saturn. Students always seem to enjoy viewing these planets.

Answers to Questions (pages 92–93)

1. b	2. a	3. d	4. c	5. b	6. c	7. d
8. b	9. a	10. d	11. c	12. d	13. b	14. c
15. a	16. d	17. b	18. c			

Answers to Mathematical Exercises (page 94–95)

Ask the students to compare their calculated answers to the values given in the text of Lesson Eight.
1. ~ 14.4 Earth masses
2. ~ 17.14 Earth masses
3. ~ 95.1 Earth masses
4. ~ 0.106 Earth masses

Lesson Eight: Asteroids, Comets, and Meteors

In the last few lessons, we have concentrated on the planets in the solar system. This is appropriate because they are the largest and most important heavenly bodies that orbit the Sun. We should know, however, that there are other bodies in the solar system.

The Asteroid Belt

Ever since astronomers have measured the distances from the planets to the Sun, they have wondered why there is such a large gap between the orbit of Mars and the orbit of Jupiter. It seems that there should be a planet inside this vast gulf. Some astronomers even constructed a number sequence (see the Math section of this Lesson) that described the orbits of the planets and suggested that another planet should exist between Mars and Jupiter.

Try as they might, however, astronomer could not find a planet in this region. Instead, after close observation, they found bodies that were much smaller than planets. They called them **minor planets** or **asteroids**.

The first few asteroids that were discovered were pretty large. **Ceres** has a diameter of about 950 km and **Pallas** is about 560 km in diameter. Others, however, were smaller, ranging from hundreds of kilometers all the way down to dust-grain size. The larger asteroids were spherical in shape, but the smaller ones were irregular. Some asteroids are rocky, but others appear to be rich in iron or carbon.

Most of the asteroids orbit in a region between about 2.1 AU and 3.3 AU, just about midway between the orbits of Mars and Jupiter. This region of the solar system is sometimes called the **asteroid belt**.

A few asteroids, however, orbit in elliptical paths that bring them inside Mars' orbit or even inside Earth's orbit. Asteroids that pass inside Mars' orbit are called **Amor asteroids**, and those that pass inside Earth's orbit are called **Apollo asteroids**.

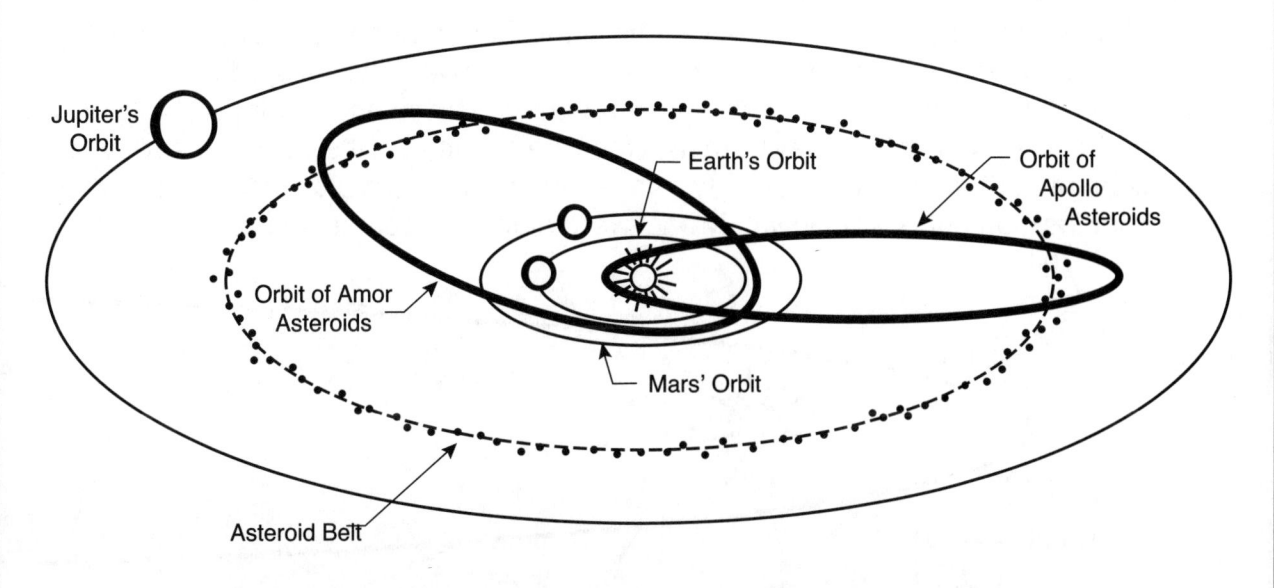

The Asteroid Belt and Amor and Apollo Asteroid Orbits

Have Asteroids Crashed Into the Earth?

Astronomers agree that the earth occasionally suffers collisions with big asteroid-sized bodies. The eroded remains of what may have been craters have been identified in Canada, Iowa, Chesapeake, and in the Yucatan Peninsula in Mexico. Some scientists feel that the collision that formed the crater in the Yucatan may have filled the earth's atmosphere with dust and caused the extinction of the dinosaurs some 65 million years ago.

Comets

Asteroids are not the only small bodies in the solar system. Occasionally, the inner portion of the solar system is visited by strange objects called comets.

A **comet** is a compact chunk of frozen gases and dust. In fact, a well-known astronomer has described comets as "dirty snowballs." Most comets are thought to be only a few kilometers in diameter. They follow long, elliptical orbits about the Sun. Most of the time, they are at large distances from the Sun, but from time to time they sweep around the narrow end of their orbits nearest the Sun. When this happens, they provide a spectacular sight for a short time.

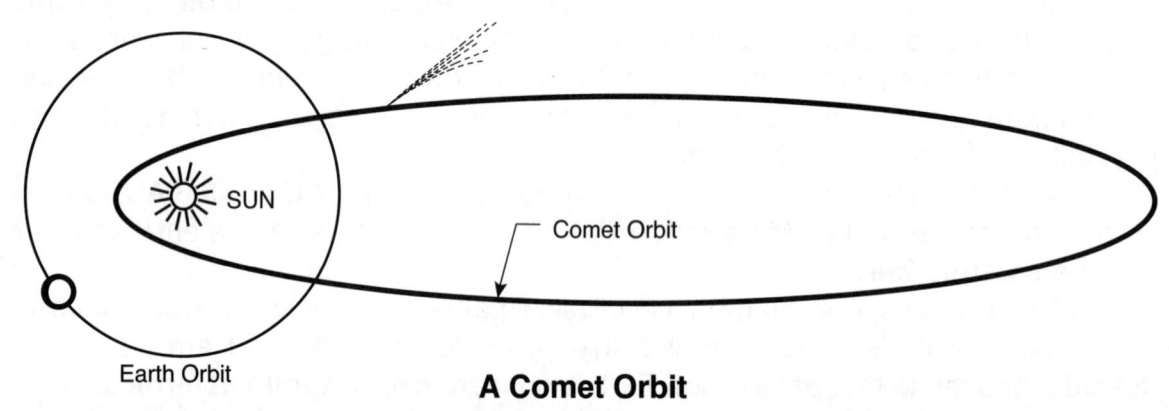

A Comet Orbit

When a comet is at large distances from the Sun, it remains totally frozen and is usually not observable. When it gets near the Sun, however, it heats up. Some of its gas is vaporized by the heat and some of its dust is released. The comet develops a big gas and dust cloud, called a **coma**, that may be many times the diameter of the earth. It also develops two tails, a **gas tail**, which points directly away from the Sun, and a **dust tail**, which points away from the Sun but is swept slightly backward along the comet's orbit.

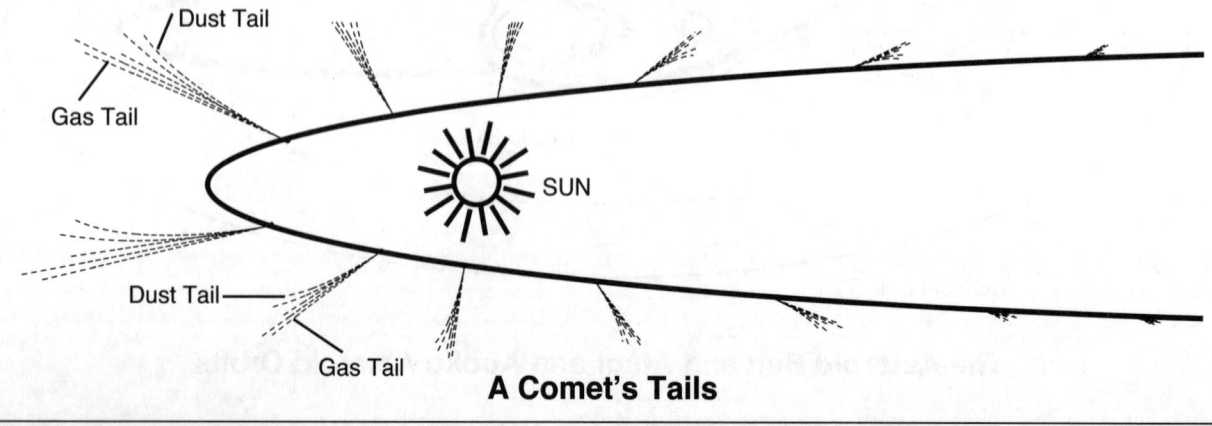

A Comet's Tails

Perhaps the most famous comet is Comet Halley that last appeared in 1986. It has an orbit period of 76 years and has appeared at many important times in history. During its 1910 visit, it was a spectacular sight, with a long tail that extended across a large portion of the sky.

Where Do Comets Come From?

Astronomers believe that the solar system is surrounded by a "cloud" of comets at distances far beyond the orbit of Pluto. Normally, they stay in this region. If something (such as the gravity of a passing star) disturbs the comet, however, it can be kicked into a long, elliptical orbit that occasionally brings it in close to the Sun. If it passes near the powerful gravity of Jupiter while it is in the inner solar system, its orbit can be further altered so that it sweeps by the Sun more often.

Meteors, Meteoroids, and Meteorites

Most of us have been outside at night and have seen flashes of light streaking across the sky. We may have called them "falling stars." What we have seen are most likely bodies entering the earth's upper atmosphere from space. As these bodies travel through the air at high speeds, they heat up the air particles and leave bright trails.

The bright trails that we see in the sky are called **meteors**. The bodies from space that cause them are called **meteoroids**. Most meteoroids are very small and completely burn up as they streak through the upper atmosphere. Some of them, however, reach the ground. Meteoroids that make it through the atmosphere and hit the ground are called **meteorites**.

Probably somewhere near 100 million meteoroids hit the earth's upper atmosphere each day. They range in size from tiny particles and grains of dust to huge chunks of rock or iron. At least two or three per day hit the earth's surface and leave meteorites.

We find several different types of meteorites. Most are small, but a few are huge, weighing several tons. Some are rocky, some are almost pure iron, and others are a mixture of rock and iron. A few contain dark, carbon material.

Where Do Meteoroids Come From?

Some of the meteoroids have their origin in the asteroid belt. Collisions between asteroids leave behind fragments of rock and iron that are sometimes flung into orbits that stray inside of Earth's orbit. When they pass close to Earth, they enter the atmosphere.

Other meteoroids, particularly the ones that come in "showers," have their origin in comets. Each time a comet passes close to the Sun, it loses a little of its material. Some of its dust gets strung out along its orbit. When the earth intersects the comet's orbit, this dust slams into the upper atmosphere. So many dust grains hit the atmosphere in a short period of time that we say it is a meteor shower. Some of the most interesting meteor showers are listed in the table on the next page. They occur during the same months each year and are named after the constellation from which the meteors seem to come. The relative directions of travel of the earth and the meteoroids make it seem like they are coming from a given region of the sky.

Dates of Annual Meteor Showers:

Shower	Approximate Date
Quadrantids	January 2–4
Delta aquarids	July 26–31
Perseids	August 10–14
Orionids	October 18–23
Geminids	December 10–13

Key Terms

Asteroid: Rocky or iron body orbiting the Sun that is smaller than a planet. Most of the asteroids move in orbits between the orbits of Mars and Jupiter.

Minor planet: An asteroid.

Ceres: One of the first asteroids to be discovered.

Pallas: One of the first asteroids to be discovered.

Asteroid belt: A region between the orbits of Mars and Jupiter extending from about 2.1 AU to about 3.3 AU where most of the asteroids orbit.

Amor asteroids: Asteroids whose orbits pass inside the orbit of Mars.

Apollo asteroids: Asteroids whose orbits pass inside the orbit of the earth.

Comet: Compact chunk of frozen gases and dust. Most comets orbit the Sun in long, elliptical orbits.

Coma: A gas and dust cloud that forms around a comet when it gets close to the Sun.

Gas tail: The tail of a comet that points directly away from the Sun.

Dust tail: The tail of a comet that points away from the Sun but is swept slightly back along the comet's orbit.

Meteor: Bright trail in the night sky caused by a body entering the earth's atmosphere.

Meteoroid: A body from space that causes a meteor.

Meteorite: A meteoroid that survives its fall through the earth's atmosphere and lands on the earth's surface.

Meteor shower: Large number of meteoroids hitting the earth's atmosphere when the earth passes through the orbit of a comet.

Name _____ Date _____

Lesson Eight—Questions

1. A rocky or iron body orbiting the Sun that is smaller than a planet is called a(n)_____. Most of these bodies move in orbits between the orbits of Mars and Jupiter.

   a) asteroid   b) minor planet   c) comet   d) both a and b

2. Two of the first asteroids to be discovered were

   a) Phobos and Deimos   b) Io and Ganymede
   c) Halley and Newton   d) Ceres and Pallas

3. The region between the orbits of Mars and Jupiter extending from about 2.1 AU to about 3.3 AU where most of the asteroids orbit is called the

   a) Oort Cloud   b) Apollo region
   c) asteroid belt   d) Amor region

4. Asteroids whose orbits pass inside the orbit of Mars are called

   a) comets   b) Amor asteroids
   c) Apollo asteroids   d) meteors

5. Asteroids whose orbits pass inside the orbit of the earth are called

   a) comets   b) Amor asteroids
   c) Apollo asteroids   d) meteors

6. A compact chunk of frozen gases and dust that orbits the Sun in a long, elliptical orbit is called a(n)

   a) comet   b) moon   c) asteroid   d) planet

7. A gas and dust cloud that forms around a comet when it gets close to the Sun is called a

   a) blanket   b) ring   c) coma   d) meteor

8. The tail of a comet that points directly away from the Sun is called its

   a) coma   b) dust tail   c) gas tail   d) ring

Lesson Eight—Questions (continued)

9. The tail of a comet that points away from the Sun but is swept slightly back along the comet's orbit is called its

    a) coma   b) dust tail   c) gas tail   d) ring

10. A bright trail in the night sky caused by a body entering the earth's atmosphere is called a

    a) meteorite   b) meteoroid   c) meteor   d) comet

11. A body from space that causes a meteor is called a

    a) moon   b) meteorite   c) crater   d) meteoroid

12. A meteoroid that survives its fall through the earth's atmosphere and lands on the earth's surface is called a

    a) meteor   b) meteorite   c) comet   d) tail

13. A large number of meteorites hitting the earth's atmosphere when the earth passes through the orbit of a comet is called a

    a) meteor shower   b) hailstorm   c) coma   d) x-ray

14. The asteroid Ceres is in an orbit that has a semimajor axis of 2.77 AU. What is the period of Ceres in years? _____

15. The asteroid Vesta has an orbit period of 3.63 years. What is the semimajor axis of its orbit? (You may want to ask your teacher to help you find a cube root for this problem.)

_____

16. Comet Encke moves in an orbit whose semimajor axis is about 2.22 AU and whose orbit eccentricity is about 0.847.
a) What is the perihelion distance of Comet Encke when it comes closest to the Sun?

_____

b) What is the aphelion distance when it is farthest from the Sun? _____

Math in Astronomy   Lesson Eight—Math: Sequences

Name _____ Date _____

Lesson Eight—Math
Sequences

Mathematicians sometimes work with collections of numbers that they call sequences. A **sequence** is a group of numbers that are arranged in some kind of order. There is a pattern to the numbers. Each number is called a **term** in the sequence. In most cases, the numbers in a sequence can be produced from a mathematical formula.

Sometimes, a sequence is based on the numbers 0, 1, 2, 3, 4, 5, … or 1, 2, 3, 4, 5, …. Something is done to these numbers to produce the terms in the sequence. Sometimes a mathematical formula is used; you plug 0, 1, 2, or some other number, into the formula and you get terms in the sequence. In other cases, there is a rule for going from one term to the next term in a sequence.

For example, the sequence: 1, 3, 5, 7, 9, … is just the odd numbers.

As another example, the sequence: 1, 4, 9, 16, 25, … is just the squares of the numbers 1, 2, 3, 4, 5, …

As yet another example, the sequence: 1, 1, 2, 3, 5, 8, 13, 21, … is a very different type of sequence. The first two numbers are defined to be 1, and each of the other numbers in the sequence is the sum of the previous two.

Exercises—Sequences

1. Suppose that each term in a sequence is produced by multiplying each of the numbers 0, 1, 2, 3, 4, 5, … by four. Write the first ten terms in the sequence.

_____

2. Suppose that each number in a sequence is produced by multiplying the square of the numbers 0, 1, 2, 3, 4, 5, … by three. Write the first five terms in the sequence.

_____

3. Suppose that the first number in a sequence is 2. The second number in the sequence is obtained by adding 1 x 3 to 2. The third number in the sequence is obtained by adding 2 x 3 to 2. The fourth number in the sequence is obtained by adding 3 x 3 to 2. The fifth number in the sequence is obtained by adding 4 x 3 to 2, and so on. Write the first eight terms of the sequence.

_____

4. Suppose that the first two terms in a sequence are 0 and 1. The third term is the sum of the first term and the second term. Each term after the third is equal to the sum of the previous two terms. Write the first eight terms of the sequence.

_____

5. Suppose that the fist two terms in a sequence are 2 and 3. The third term and all terms thereafter are obtained by multiplying the previous two terms together. Write the first six terms of the sequence.

_____

© Mark Twain Media, Inc., Publishers

Math in Astronomy   Lesson Eight—Math: The Titius-Bode Rule

Name _____ Date _____

The Titius-Bode Rule

An interesting sequence in astronomy is called the **Titius-Bode Rule**. It was dreamed up in 1766 when a German astronomer, Johann Titius, found that he could generate a sequence that approximately predicted the semimajor axes of the planets' orbits in AU. His method was made popular by a better-known astronomer called Johann Bode, so the sequence is called the Titius-Bode Rule. The rule goes like this:

1. Write down 0 as the initial value of the first term.
2. Then write down 3 as the second term.
3. Generate higher terms by multiplying the previous term (starting with the second one) by 2. At this point, you should have the sequence:

    0   3   6   12   24   48   96   192

4. Now, add four to each of the terms. You should have the sequence:

    4   7   10   16   28   52   100   196

5. Finally, divide each term by 10. You should end up with the following sequence:

    0.4   0.7   1.0   1.6   2.8   5.2   10.0   19.6

Titius and Bode compared their values to the actual distances of the planets (known at that time) from the Sun (Uranus was added in 1781) and found that their sequence (approximately) predicted the planet orbits, except for the 2.8 value.

Planet	Semi-major axis	Titius-Bode value
Mercury	0.38	0.40
Venus	0.72	0.70
Earth	1.00	1.00
Mars	1.52	1.60
		2.80
Jupiter	5.20	5.20
Saturn	9.54	10.00
Uranus	19.18	19.60

With the exception of the 2.80 value, this sequence fits (approximately) the average distances of the planets from the Sun (out to Uranus.) Astronomers of the day felt that there was something important about the way that the Titius-Bode sequence "predicted" the positions of the planets. They thought that the 2.80 value indicated an unknown planet that astronomers had not been clever enough or thorough enough to discover.

Exercise—Titius-Bode Rule

6. What is the next number in the Titius-Bode sequence after the last one in the table above? _____ Does it predict the position of Neptune or Pluto? _____

Math in Astronomy   Lesson Eight—Teacher's Page

Lesson Eight—Teacher's Page

Class Activities

I. The Barringer Meteorite crater, near Winslow, Arizona, is a relatively recent feature. Scientists estimate that it was formed as recently as 50,000 years ago when a meteorite a little less than 100 km in diameter crashed into the earth. The crater, about a mile in diameter, is well defined and can be visited. It has been estimated that the energy released by the impact of the meteorite would have been equivalent to the explosion of a three-megaton hydrogen bomb.

The prospect of a collision between the earth and a large meteorite can be scary. Some astronomers have performed simulation calculations with computers that indicate that a "nuclear winter" (similar to what might happen after an all-out nuclear war) might occur if enough particulate material were blasted into the upper atmosphere by the impact of a large meteorite on the earth. Crops might not grow and temperatures could be depressed for several seasons. Some scientists feel that a large impact of this type may have led to the extinction of the dinosaurs about 65 million years ago.

It might be interesting to obtain a picture of the Barringer crater and use it as a lead-in to a discussion of what might happen if a very large meteorite hit the earth some time in the near future. How would we as humans cope and what spirit of cooperation might have to emerge among diverse cultures in order for us all to survive? How might other species (plant and animal) fare?

II. Watching a meteor shower can be a lot of fun if the temperature is not too cold and the observing site is comfortable. Counting the number of meteors seen per hour is a typical activity. If students indicate that they would like to watch a meteor shower, you might suggest that they find a site in an open area with an unobstructed view of the horizon, and recline in a lawn chair. This will give them a view of most of the sky. A blanket and a cup of hot chocolate might make the experience a little better if the weather is even a little cool.

Choosing a viewing time after midnight will usually show more meteors. After midnight, the viewer's position on the earth will be on the side that is in the direction of the earth's orbital motion, resulting in a higher relative velocity between the viewer and the meteoroids.

Parental supervision is, of course, advised for any nighttime activity.

Answers to Questions (pages 101–102

1. a 2. d 3. c 4. b 5. c 6. a 7. c 8. c 9. b 10. c 11. d 12. b 13. a
14. 4.61 years   15. 2.36 AU   16a. 0.34 AU   16b. 4.10 AU

Answers to Math Exercises (pages 103–104)

1. 0, 4, 8, 12, 16, 20, 24, 28, 32, 36
3. 2, 5, 8, 11, 14, 17, 20, 23
5. 2, 3, 6, 18, 108, 1944

2. 0, 3, 12, 27, 48
4. 0, 1, 1, 2, 3, 5, 8, 13

6. 38.8; Pluto

© Mark Twain Media, Inc., Publishers

Lesson Nine: The Sun and the Stars

If you are outside on a clear night in a place with very few lights, you can see as many as 3,000 stars with just your eyes, and many times that number if you use binoculars or a small telescope. The stars are everywhere in the sky and, on a moonless night, even provide us with a little starlight so that we can see dimly. Unlike planets and moons, which shine by reflected light, stars emit their own light.

Our Sun

Although we usually think of our Sun as being something very special, it is really just an ordinary star. Compared to the rest of the solar system, however, it **is** something special. The Sun contains about 98% of the mass of the solar system and provides almost all of its energy. Its diameter is about 110 times the diameter of the Earth.

The center portion or **core** of the Sun is a blazing furnace where hydrogen is being converted into helium. The conversion process is the same one that goes on in a hydrogen bomb, and the temperature in the core is a toasty 15 million K! Several layers above the core transport the energy upward toward the surface of the Sun, the part we call the **photosphere.**

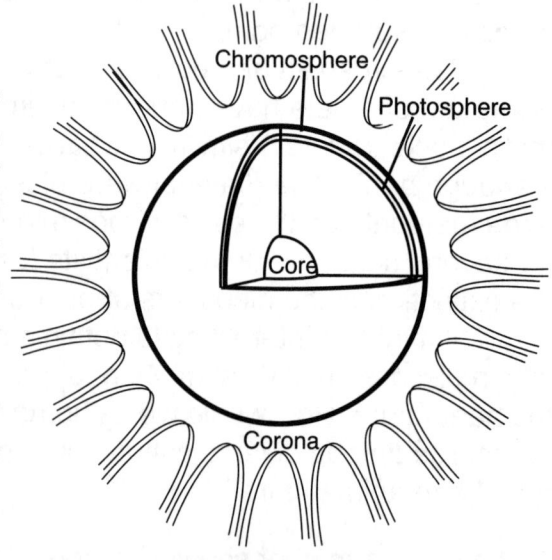

The Sun

The photosphere is the bright disk that we see when we look towards the Sun. We might think of it as the "surface" of the Sun. It is only about 500 km thick and its temperature is about 6,000 K, much cooler than the core. Upward currents of hot gases carry energy up to it from lower layers. Outer layers of the Sun above the photosphere are called the **chromosphere** and the **corona.**

The Stars

It seems like there are a lot of stars in the sky when we look at the sky on a dark night, but we can only see a few thousand at most. The number of stars that we can see with a high powered telescope is almost unbelievably large, billions and billions. They are obviously a popular item in the universe.

When you first look at the sky with just your eyes, the stars seem pretty much alike. If you pay closer attention, however, you will see that some stars are brighter than others and that not all of the stars are white. Some of them are bluish, yellowish, or even reddish in color. All of them have a few things in common, however. Like our Sun, they are large, they are very hot, and they shine by their own light.

There are many types of stars, and astronomers have seen fit to classify them in many ways. For our purposes, however, three types will be sufficient.

Ordinary stars are stable stars that shine with a steady light. Larger ones are hotter and may be blue, blue-white, or white in color. Smaller ones are cooler and may be white, yellow, orange, or even red.

Red giant stars are huge, bright stars that are many times the diameter of the Sun. Often they are ordinary stars that have swollen to a large size. Their surfaces are very cool, however, only about 3,000 K, and they shine with a reddish light. Their brightness comes from their large size.

White dwarf stars are very small stars that glow white-hot. Some of them are not much larger than the earth. They are very dense and give off a lot of light for their size.

Binary Stars

Not all of the stars are by themselves in the sky. More than a few of them are in pairs. They are called **binary stars.** Binary stars are held together by their own gravity and revolve in orbits around a central point called their "center of mass."

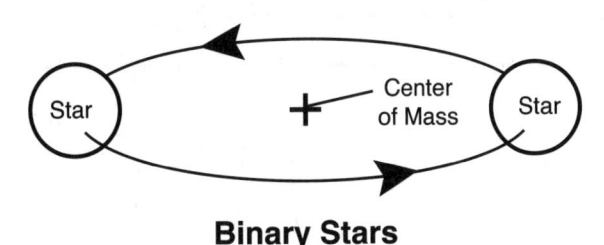

Binary Stars

Other stars exist in clusters of three or more.

Light From the Stars

So far, we have not been able to travel to the stars and measure their properties firsthand. What we know about them comes from observations of the stars and measurements made of the starlight that reaches the earth.

We are all familiar with light. We know that it is difficult to see when there is not enough light, and that we can turn on a light switch to flood a dark room with light. What we often don't know is that the visible light that we can see with our eyes is only a small part of the total amount that is present.

Light is an **electromagnetic wave.** We can think of an electromagnetic wave as a wavy volume of energy that can travel through space at a very high speed. Our most familiar example of electromagnetic waves is the visible light that illuminates our world. Other examples include radio waves, microwaves, infrared light, ultraviolet light, x-rays, and even gamma rays. Infrared light is not visible to us, but it gives us a sensation of warmth when it shines on our skin. Ultraviolet light also is not visible to us, but it can give us suntans and sunburns.

We sometimes draw pictures of electromagnetic waves like the one on the next page. The waves of energy are drawn as waves like we might see on the surface of a pool of water. The high points on the waves are called the peaks and the low points are called the troughs.

How high the peaks are or how low the troughs are is a measure of how strong the wave is or how bright the light would appear if we could see it.

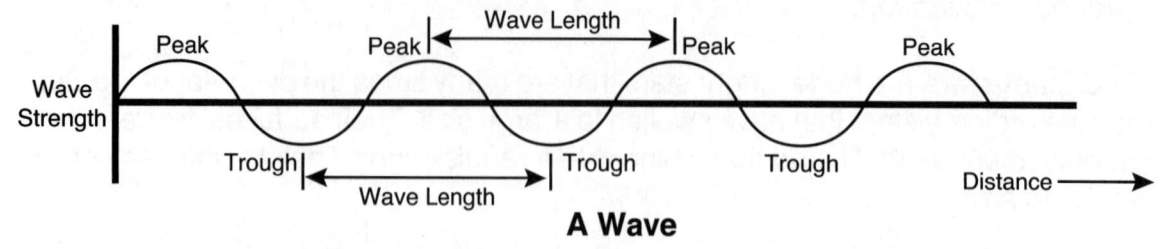

A Wave

Of more interest to us, however, is the distance between two peaks or two troughs that are next to each other. This distance is called the **wavelength** of the wave. The wavelength of an electromagnetic wave determines which type of wave it will be. For example, ones with wavelengths many meters or kilometers long are radio waves, and ones with wavelengths about a centimeter long are called microwaves. Ones with shorter wavelengths are called infrared light. Visible light and ultraviolet light have even shorter wavelengths.

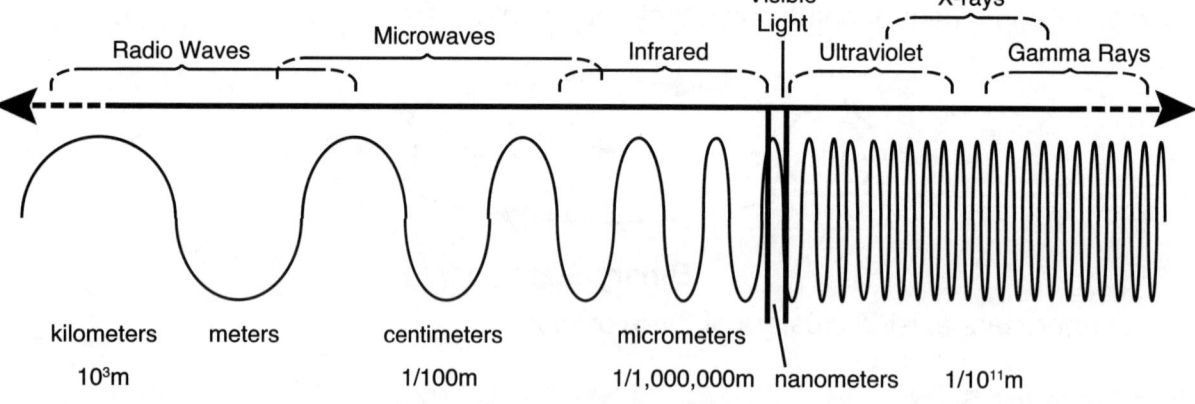

Electromagnetic Waves

It is convenient to measure the wavelengths of visible light in units called **nanometers**. A nanometer (nm) is one billionth of a meter. The colors of visible light that we see and their wavelengths in nanometers are shown below.

Color	Wavelength range
Violet	400 nm to 450 nm
Blue	450 nm to 500 nm
Green	500 nm to 550 nm
Yellow	550 nm to 600 nm
Orange	600 nm to 650 nm
Red	650 nm to 700 nm

Astronomers measure the light from stars in many wavelength ranges. We are familiar with ordinary telescopes that allow us to view the visible light from the stars. There are also radio telescopes that look at radio waves from the stars, infrared telescopes that see stars in infrared wavelengths, ultraviolet telescopes that view the stars in ultraviolet light, and even x-ray and gamma ray telescopes.

Distances to the Stars

Even the closest stars are at huge distances from the solar system. To try to describe their distances with ordinary units such as miles or kilometers would force us to use very large numbers. Even distances in Astronomical Units would be pretty big. We choose to use a much larger unit to measure the distance to the stars, a unit called a **light year.** It is based on the speed of light.

We know that light of all wavelengths travels at a constant speed through space. The speed of light can be expressed in many different units. Some of them (approximate values) are shown below.

Speed of light = 186,000 miles/second
= 300,000 kilometers/second
= 3.00×10^8 meters/second
= 0.0020 AU/second

A light year is the distance that light would travel in a year at the speed shown above. Its value in kilometers and AU is:

1 light year (ly) = 9.46×10^{12} km
= 63,240 AU

Even the closest stars are several light years away from the solar system. The distances to the three closest stars and the distances to some of the brightest stars are shown in the tables below.

Distances to the three nearest stars:

Star	Distance
Alpha Centauri	4.3 ly
Barnard's Star	5.9 ly
Wolf 359	7.6 ly

Distances to the brightest stars (as seen from Earth in the Northern Hemisphere):

Star	Distance	Star	Distance
Sirius	8.6 ly	Arcturus	36.0 ly
Vega	26.5 ly	Capella	45.0 ly
Rigel	900.0 ly	Procyon	11.3 ly
Betelgeuse	520.0 ly	Altair	16.5 ly
Aldebaran	68.0 ly	Spica	220.0 ly
Antares	520.0 ly	Fomalhaut	22.6 ly
Pollux	35.0 ly	Deneb	1,600.0 ly

Lives of Stars

When we look at the stars, they seem pretty much the same every night. However, astronomers have been able to determine that, like most things, stars have a lifetime. They are born, they exist for a time, and then they die. Very large, hot stars do not live for very long, maybe only a few hundred million years. Smaller, cooler stars live longer, sometimes for several billions of years.

Stars are formed from giant clouds of gas and dust that are sometimes called **nebulae** (the plural of nebula). The gas in a nebula is mainly hydrogen. Clumps of gas and dust develop in the nebula and, because of their own gravity, begin to contract. As they contract, they become warmer. The larger ones get hot enough to become stars, and the smaller ones become planets.

A large clump goes through several stages on the way to becoming a star. First it contracts and gets warmer and warmer. Eventually it becomes so hot that it begins to glow red. As it contracts more and becomes even hotter, its interior gets warm enough that hydrogen starts to be converted into helium. This is the same nuclear reaction that occurs in the hydrogen bomb. When the nuclear reaction settles in, the clump has become a full-fledged star.

A star spends most of its life converting hydrogen into helium and emitting light at a more or less constant rate. Toward the end of its life, however, it has used up most of its hydrogen. It swells up into a red giant for a while, and then shrinks down again and begins to convert helium into things like carbon and oxygen. If it is a smaller star, this is its last phase. When it has used up most of its helium, it swells up again, throws off its outer layers and exposes its white-hot core of carbon and oxygen. The star has become a white dwarf. It ends its life by slowly cooling into a small reddish star and then into a black cinder.

If the star is large, it may go through several other phases and then eventually explode in a huge detonation called a **supernova.** A supernova is the death of a large star. After the explosion, a small, incredibly dense body called a **neutron star** may be left. The material in a neutron star is so dense that a teaspoon full of it would weigh millions of tons.

If the star is very, very large, it may leave behind an even stranger body after the supernova, a **black hole.** A black hole is a body that is so dense and heavy that its own gravity keeps on contracting it until it is just a tiny dot. It is a tiny dot, however, with an unimaginably large density. Anything that gets too close to the black hole will be sucked in and will remain there. Nothing, not even light, can get out.

Key Terms

Core: The innermost region of the Sun where the conversion of hydrogen to helium is taking place.

Photosphere: The bright disk that we see when we look toward the Sun.

Chromosphere: A bright, gaseous area just above the photosphere on the Sun.

Corona: The very thin, outer layer of the sun's atmosphere. It is very hot but very low-density.

Ordinary star: A stable star that shines with a steady light. Stars in this category are converting hydrogen into helium in their cores. Most stars that we see in the sky are in this category.

Red giant star: A huge, bright star that is many times the diameter of the Sun. Stars in this category have surfaces that are very cool. They shine with a reddish light.

White dwarf star: Very small, white-hot stars that are about the size of the earth.

Binary star: A pair of stars that are held together by their own gravity and revolve in orbits around a central point.

Electromagnetic wave: A wavy volume of energy that travels through space at a high speed.

Wavelength: The distance between two adjacent peaks or two adjacent troughs of an electromagnetic wave.

Nanometer: One billionth of a meter. The wavelengths of visible light are measured in nanometers.

Light year: The distance that light travels through space in one year. Equal to 63,240 AU.

Nebula: A cloud of gas and dust in space. It appears as a small, hazy patch of light in the sky when viewed from Earth. (The plural of nebula is nebulae.)

Supernova: The death of a star with a large mass. It is a huge explosion that releases large amounts of energy.

Neutron star: The remains of a large star after a supernova. It is very small (only a few kilometers in diameter), but incredibly dense. Stars whose remains have masses between about 1.4 and 3 times the mass of the Sun can become neutron stars.

Black hole: The remains of a very large star after a supernova. All of the mass of the star shrinks down to a tiny point. Stars whose remains are larger than about three times the mass of the Sun can become black holes.

Name _____ Date _____

Lesson Nine—Questions

1. The innermost region of the Sun where the conversion of hydrogen to helium is taking place is called the

    a) photosphere   b) core   c) corona   d) chromosphere

2. The bright disk that we see when we look toward the Sun is called the

    a) photosphere   b) core   c) corona   d) chromosphere

3. The bright, gas area just above the photosphere on the Sun is called the

    a) photosphere   b) core   c) corona   d) chromosphere

4. The very thin, outer layer of the sun's atmosphere is called the _____ . It is very hot but very low-density.

    a) photosphere   b) core   c) corona   d) chromosphere

5. A stable star that is converting hydrogen to helium and shines with a steady light is called a(n)

    a) ordinary star   b) black hole   c) red giant   d) white dwarf

6. A huge, bright star, many times the diameter of the Sun, with a cool surface that shines with a reddish light is called a(n)

    a) ordinary star   b) black hole   c) red giant   d) white dwarf

7. A very small, white-hot star that is about the size of the Earth is called a(n)

    a) ordinary star   b) black hole   c) red giant   d) white dwarf

8. A pair of stars that are held together by their own gravity and revolve in orbits around a central point are called

    a) red giant stars   b) white dwarf stars   c) binary stars   d) neutron stars

9. The distance between two adjacent peaks or two adjacent troughs of an electromagnetic wave is called its

    a) infrared   b) wavelength   c) speed   d) light year

Name _____ Date _____

Lesson Nine—Questions (continued)

10. Wavelengths of visible light are measured in

    a) kilograms   b) Kelvins   c) Hertz   d) nanometers

11. The distance unit that we use for measuring the distances to the stars is called a(n) _____. It is the distance that light travels through space in one year.

    a) kilometer   b) astronomical unit   c) light year   d) nanometer

12. A cloud of gas and dust in space that appears as a small, hazy patch of light when viewed from Earth is called a

    a) supernova   b) nebula   c) chromosphere   d) black hole

13. The huge explosion that is the death of a massive star is called a

    a) supernova   b) nebula   c) big bang   d) light year

14. The remains of a large star after a supernova that is very small (only a few kilometers in diameter) but very dense is called a(n)

    a) red giant   b) black hole   c) neutron star   d) asteroid

15. The remains of a very large star after a supernova whose mass has contracted down to a tiny point is called a

    a) red giant   b) black hole   c) neutron star   d) white dwarf

16. Visible light with a wavelength of 560 nm is

    a) red   b) blue   c) yellow   d) green

17. Visible light with a wavelength of 440 nm is

    a) violet   b) blue   c) orange   d) green

18. Visible light with a wavelength of 680 nm is

    a) red   b) orange   c) yellow   d) violet

Math in Astronomy   Lesson Nine—Questions

Name _____ Date _____

Lesson Nine—Questions (continued)

19. The speed of light is

    a) 300,000 m/sec   b) 300,000 miles/sec
    c) 300,000 km/sec   d) 300,000 AU/sec

20. The speed of light is

    a) 100,000 mi/sec   b) 186,000 mi/hour
    c) 286 km/sec   d) 186,000 mi/sec

21. A light year is

    a) 9.46×10^{12} AU   b) 300,000 km
    c) 63,240 AU   d) 186,000 miles

22. Which of the following stars is closest to the solar system?

    a) Sirius   b) Barnard's star   c) Procyon   d) Capella

23. Which of the following stars is closest to the solar system?

    a) Sirius   b) Arcturus   c) Procyon   d) Betelgeuse

24. Which of the following stars is farthest from the solar system?

    a) Altair   b) Rigel   c) Procyon   d) Capella

25. Which of the following stars is farthest from the solar system?

    a) Rigel   b) Deneb   c) Betelgeuse   d) Spica

Math in Astronomy   Lesson Nine—Math: Distances in Light Years/Inverse Square Law of Radiation

Name _____   Date _____

Lesson Nine—Math

Distances in Light Years

As was mentioned in the lesson, distances to the stars are so large that they are often measured in light years. A light year (ly) is equal to about 63,240 AU. You can convert between light years and Astronomical Units by the following conversion equations.

Number of AU = number of ly x 63,240 AU/ly

or

Number of ly = number of AU x 0.00001581 ly/AU

Exercises—ly-AU Conversion

1. How many ly is 39.44 AU? _____

2. How many ly is 20,000 AU? _____

3. How many ly is 100,000 AU? _____

4. How many AU is 0.1234 ly? _____

5. How many AU is 4.30 ly? _____

6. How many AU is 0.0105 ly? _____

The Inverse Square Law of Radiation

Stars that appear bright to us can either be very bright or very close, and stars that appear dim can either be very dim or very far away. Far-away stars appear dimmer than nearby ones because the light that they emit is spread out over the vast distances of space before it reaches us. To see how this happens, look at the diagram shown below.

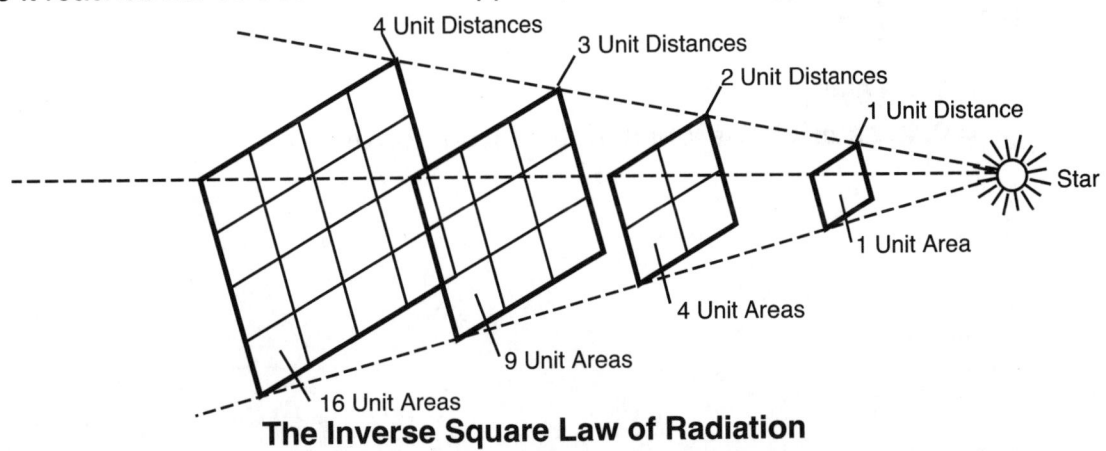

The Inverse Square Law of Radiation

© Mark Twain Media, Inc., Publishers   115

As a portion of the light travels away from the star, it covers the small square at one distance unit from the star. At two distance units from the star, that same amount of light would spread out to cover four of the squares that it covered when it was one unit away. At a larger distance of three units, the light would have spread out even more and would cover nine of the squares. At the even larger distance of four units, the light would be so spread that it would cover 16 squares. Notice that the number of squares that the light covers is related to the square of the number of distance units from the source.

The same amount of light keeps spreading out as it gets farther and farther from its source. Therefore, as you get farther from the source, there is less light intensity or brightness at any one spot. We can express this mathematically as follows:

$$\frac{\text{Brightness at spot 1}}{\text{Brightness at spot 2}} = \left(\frac{\text{distance of spot 2 from light source}}{\text{distance of spot 1 from light source}}\right)^2$$

or, in another form:

$$\text{Brightness at spot 1} = \text{Brightness at spot 2} \times \left(\frac{\text{distance of spot 2 from light source}}{\text{distance of spot 1 from light source}}\right)^2$$

The distances can be in any units that we choose, as long as they are both in the same units. We will express brightness as some number of times as bright as the sun.

Let's see how we can use this little piece of mathematics. Suppose a certain star appears 2 times as bright as the Sun when it is viewed from a distance of 0.5 light years. How bright would it appear to an observer 10 light years away? We will call the place that is 0.5 light years away **spot 2**, and the place that is 10 light years away **spot 1**.

$$\text{Brightness} = 2 \times \left(\frac{0.5}{10}\right)^2$$

$$= 2 \times 0.05^2$$
$$= 0.005 \text{ times as bright as the Sun}$$

As another example, consider how bright the Sun appears to someone viewing it from Mars as compared to someone viewing it from Earth. Mars is at 1.52 AU from the Sun and Earth is at 1 AU. We will consider the brightness of the Sun as viewed from Earth to be 1 brightness unit. The Earth in this case will be **spot 2** and Mars will be **spot 1.**

$$\text{Brightness} = 1 \times \left(\frac{1}{1.52}\right)^2$$

$$= 1 \times 0.658^2$$
$$= 0.43 \text{ as bright as when viewed from Earth}$$

Exercises—Inverse Square Law of Radiation

7. A certain star appears 4 times as bright as the sun when viewed from a distance of 1 light year. How bright would it appear when seen from a distance of 4 light years?

8. A certain star appears 10 times as bright as the Sun when viewed from a distance of 2 light years. How bright would it appear if viewed from a distance of 1 light year?

9. How many times as bright would the Sun appear to someone viewing it from Venus (at 0.72 AU) as compared to someone viewing it from Earth (at 1 AU)?

10. How many times as bright would the Sun appear to someone viewing it from Jupiter (at 5.20 AU) as compared to someone viewing it from Earth (at 1 AU)?

11. How many times as bright would the Sun appear to someone viewing it from Pluto (at 39.4 AU) as compared to someone viewing it from Earth (at 1 AU)?

Math in Astronomy   Lesson Nine—Teacher's Page

Lesson Nine—Teacher's Page

Comments

The **ordinary stars** that we discuss in this lesson are called **main sequence stars** by astronomers because of their plotted position on a type of graph called a Hertzsprung-Russell diagram.

Class Activities

I. Some of the stars are at relatively small distances from the solar system. For example, the star Sirius is 8.6 light years from us. The light from that star takes 8.6 years to reach the Earth. It is sometimes interesting for students to estimate how old they were and what they were doing when the light left the star. They might want to do this for Procyon (11.2 light years away), Barnard's star (6.0 light years away), or Alpha Centauri (4.3 light years away) also.

II. Other stars are at vast distances from the Earth. Students might have fun speculating about what historical events were taking place on Earth when the light that we see today left a given star. Some ones used might be Capella (43 light years away), Aldebaran (55 light years away), Spica (228 light years away), Betelgeuse (522 light years away), or Deneb (1,600 light years away).

Answers to Questions (pages 112–114)

1. b 2. a 3. d 4. c 5. a 6. c 7. d 8. c 9. b 10. d 11. c 12. b 13. a
14. c 15. b 16. c 17. a 18. a 19. c 20. d 21. c 22. b 23. a 24. b 25. b

Answers to Math Exercises (pages 115–117)

1. 0.0006256 ly   2. 0.316 ly   3. 1.58 ly
4. 7,804 AU   5. ~ 272,000 AU   6. 664 AU

7. 0.25 times as bright as the Sun
8. 40 times as bright as the Sun
9. 1.93 as bright as when viewed from Earth
10. 0.037 as bright as when viewed from Earth
11. 0.00064 as bright as when viewed from Earth

Lesson Ten: The Milky Way and Other Galaxies

Stars are not randomly scattered across the universe. Instead, we find them clustered in groups that we call galaxies. Galaxies can contain billions or even hundreds of billions of stars. Some of them are giant spirals, some of them are shaped like ellipses, and others are irregular in shape.

The Milky Way Galaxy

We live in a large, spiral galaxy called the Milky Way. If we could see it from far out in space it would look something like the pinwheel shape shown in the diagram at right. Our Galaxy is about 100,000 light years in diameter. Most of its outer region is a flattened spiral of stars called the **Galactic disk.** The stars in the Galactic disk are organized into three or four spiral trails, called **arms,** that wind outward from the center. The Sun is somewhere between 25,000 and 30,000 light years from the center of the galaxy along one of these spiral arms. The galactic disk is relatively thin, only about 2,000 light years thick.

Milky Way Galaxy

The center of the galaxy is filled by a large sphere of closely packed stars called the **central bulge.** It is about 25,000 light years in diameter. Its shape is more apparent in the edge-on view of the galaxy shown below.

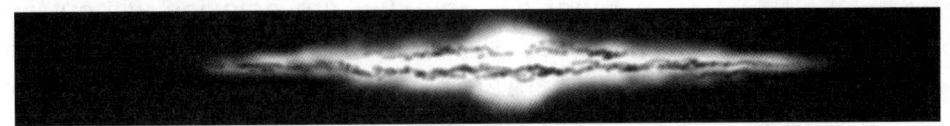

Milky Way Galaxy Edge-On View

The Milky Way is home to a huge number of stars, about 100 billion, and has an estimated mass of around 200 billion times the mass of the Sun. These numbers are so large that they almost overwhelm a person's imagination. The galaxy also contains vast clouds of dust and gas. The dust clouds make it difficult for us to see for long distances through the galactic disk and the central bulge.

Our galaxy is surrounded by a spherical cloud of more than 200 clusters of stars. The cloud is called the **halo** and the clusters of stars are known as **globular clusters.** The globular clusters are around 30 to 100 light years in diameter, and contain 100,000 to 1,000,000 stars each.

When we look at the sky at night, we see the Milky Way as a whitish-gray swath across the heavens. This is our view from Earth of the galactic disk and the central bulge. We only see a part of these because much of our view is obscured by dust clouds. If we look in the direction of the constellation Sagittarius, we are looking toward the center of the galaxy, and if we look toward Perseus, we are looking outward towards the edge of the galactic disk.

Our Neighbor Galaxies

Our galaxy resides in a region of space that contains about 36 galaxies that are clustered together in what astronomers call the **local group.** Most of the galaxies in the group are smaller than the Milky Way and do not have its spiral shape. Only one other, the Andromeda Galaxy, is a large spiral.

Even though the galaxies in the local group are said to be clustered together, the distances that separate them are huge. Close neighbors of the Milky Way, the Larger Magellanic Cloud and Smaller Magellanic Cloud galaxies, are about 200,000 light years from us, and the Andromeda Galaxy is a distant 2.5 million light years away.

Other Galaxies

Are there other galaxies out there beyond the local group, and if so, do we know how far away they are? The answer to both of these questions is yes. There are billions of other galaxies out there and the distances to them are almost unbelievably large.

Galaxies come in several shapes and sizes. Scientists have divided them into three basic types: **spiral galaxies, elliptical galaxies,** and **irregular galaxies.**

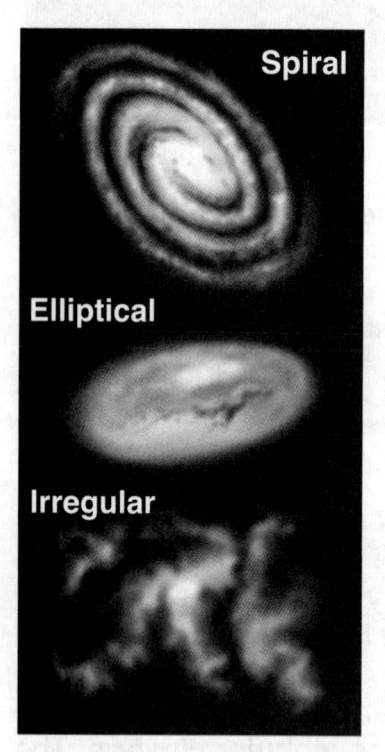

Spiral galaxies are pinwheel-shaped galaxies like our own Milky Way. They have a flat disk-like part that is filled with stars, gas, and dust. The stars wind outward from the center in spiral arms. Sometimes there is a spherical bulge in the center of them.

The dust and gas clouds in the spiral arms of this type of galaxy provide a good place for new stars to form. There are many young stars in the disk portion of spiral galaxies.

Elliptical galaxies are galaxies without spiral arms. They are shaped like ellipsoids (three-dimensional ellipses) or spheres. There are more elliptical galaxies than any other type.

Elliptical galaxies do not appear to contain much gas or dust, and most of the stars in them appear old. New stars do not seem to form easily in galaxies without a lot of gas and dust clouds.

Irregular galaxies are small galaxies that do not have a definite shape. Some of them just look like big clusters of stars, while others have some structures that look like unattached pieces of spiral arms. There is a lot of dust and gas as well as many young stars in irregular galaxies.

Astronomers use many methods to find the distances to galaxies. The distances to galaxies in the local group and to other galaxies out to about 100,000,000 light years can be found from the properties of stars, called Cepheid variable stars, whose brightnesses change with regular periods. For distances greater than this, less precise methods must be used. Galaxies more than 10 billion light years away have been found.

The Motion of the Galaxies

When astronomers began to study galaxies in detail, they were surprised to learn that almost all of the other galaxies are moving away from the Milky Way. The farther away a galaxy was from the Milky Way, the faster that it seemed to be moving away from us. As they learned more, they realized that the galaxies were also moving away from one another. This led them to believe that the universe is expanding.

The Origin of the Universe

The fact that all of the galaxies seemed to be moving away from one another caused astronomers to wonder if, at some time, they were not all in one spot. Perhaps a giant explosion sent them all rushing out away from each other. This sort of thought plus quite a few other considerations have led astronomers to construct a theory for the origin of the universe called the **big bang.** It is a lot more complicated than just an explosion that blasted galaxies out into space. We will look at a very simplified view of it.

In the beginning, the universe was just a tiny spot. Matter, energy, time, and even space did not exist as we know them now. Then, in an explosion known as the big bang, the universe came into being. Vast amounts of energy and space itself expanded outward after the big bang. It would be wrong, however, to say that they expanded from any one point. Since space was created as part of the big bang, we have to think of the big bang as occurring more or less everywhere. Things can expand with respect to each other but not relative to any fixed point.

Initially, the universe was very hot, a fireball of radiation similar in some ways to the electromagnetic waves that we talked about in Lesson Nine. It would be much later, when the universe had expanded and cooled, that the galaxies and stars would form.

At present, the universe is still expanding. Whether the universe will continue to expand, or whether the expansion will cease and the universe will contract again, is still being debated and investigated by scientists.

Key Terms

Galactic disk: Flattened spiral of stars that forms the outer portion of the Milky Way and other spiral galaxies. It contains stars, gas, and dust, and is rich in new, young stars.

Arm: Spiral trail of stars that winds outward from the center of the Milky Way and other spiral galaxies.

Central bulge: Large sphere of closely packed stars at the center of the Milky Way Galaxy.

Halo: Spherical cloud of more than 200 clusters of stars that surrounds the Milky Way Galaxy.

Globular cluster: Spherical clusters of stars, around 30 to 100 light years in diameter, found in the halo of the Milky Way.

Local group: Cluster of about 36 galaxies that includes the Milky Way and its nearest neighbor galaxies.

Spiral galaxy: Pinwheel-shaped galaxies like our own Milky Way. They have a flat disk-like part that is filled with stars, gas, and dust. The stars wind outward from the center in spiral arms. Sometimes there is a spherical bulge in the center.

Elliptical galaxy: Galaxies without spiral arms. They are shaped like ellipsoids or spheres.

Irregular galaxy: Small galaxies that do not have a well-defined shape.

Big bang: A theory of the origin of the universe, in which everything—matter, energy, space, and so forth—expands from a single point.

Name _____ Date _____

Lesson Ten—Questions

1. The flattened spiral of stars that forms the outer portion of the Milky Way and other spiral galaxies is called the _____ . It contains stars, gas, and dust, and is rich in new, young stars.

    a) central bulge   b) halo   c) arm   d) galactic disk

2. A spiral trail of stars that winds outward from the center of the Milky Way and other spiral galaxies is called a(n)

    a) central bulge   b) halo   c) arm   d) galactic disk

3. The large sphere of closely packed stars at the center of the Milky Way Galaxy is called the

    a) central bulge   b) halo   c) arm   d) galactic disk

4. The spherical cloud of more than 200 clusters of stars that surrounds the Milky Way Galaxy is called the

    a) central bulge   b) halo   c) arm   d) galactic disk

5. Spherical clusters of stars, around 30 to 100 light years in diameter, found in the halo of the Milky Way, are called

    a) local groups   b) halos   c) globular clusters   d) central bulges

6. A cluster of about 36 galaxies that includes the Milky Way and its nearest neighbor galaxies is called the

    a) halo   b) local group   c) galactic disk   d) arm

7. A type of galaxy that is pinwheel-shaped like our own Milky Way is a(n)

    a) elliptical galaxy   b) spiral galaxy   c) irregular galaxy

8. A type of galaxy without spiral arms that is shaped like an ellipsoid or a sphere is a(n)

    a) elliptical galaxy   b) spiral galaxy   c) irregular galaxy

9. A small galaxy that does not have a well-defined shape is a(n)

    a) elliptical galaxy   b) spiral galaxy   c) irregular galaxy

10. A theory of the origin of the universe, in which everything—matter, energy, space, and so forth—expands from a single point is called the

    a) redshift   b) genesis   c) powerball   d) big bang

Math in Astronomy   Lesson Ten—Math: Hubble's Law

Name _____ Date _____

Lesson Ten—Math

Hubble's Law

In this lesson, we mentioned that the galaxies appear to be moving away from the earth and away from each other. The speed at which a galaxy appears to be moving away from the earth is called its **radial velocity** by astronomers. We also said that the farther away a galaxy was, the faster it seemed to be moving away from us, i.e. the larger its radial velocity.

An astronomer, Edwin Hubble, found that there is a mathematical relation between how far away a galaxy is and how fast it is moving away from us. The relation is called **Hubble's Law** and involves a number called the **Hubble constant.**

$$\text{distance to galaxy} = \frac{\text{radial velocity of galaxy}}{\text{Hubble constant}}$$

The radial velocity must be in units of km/second and the distance must be in millions of light years. Astronomers have a number of methods for determining how far away a galaxy is. They also have a good method for determining radial velocities of galaxies that we will look at in the next section. Using many measurements of distances to galaxies and their radial velocities, astronomers have come up with values for the Hubble constant. There are several estimates for the Hubble constant. We will use 20 km/sec per million light years.

Sometimes the distance to a galaxy cannot be determined easily. In that case, we can measure the radial velocity of a galaxy and use Hubble's Law to determine the distance to it.

For example, suppose that the radial velocity of a certain galaxy is found to be 1,000 km/sec. What is the distance of the galaxy from the earth?

$$\text{distance} = \frac{1,000 \text{ km/sec}}{20 \text{ km/sec per million light years}}$$

$$= \frac{1,000}{20} \text{ million light years}$$

$$= 50 \text{ million light years}$$

Exercises—Hubble's Law

1. If the radial velocity of a galaxy is 2,000 km/sec, what is the distance to the galaxy?
_____

2. If the radial velocity of a galaxy is 4,000 km/sec, what is the distance to the galaxy?
_____

3. The radial velocity of a galaxy in the constellation Virgo is about 1,200 km/sec. What is the distance to the galaxy? _____

4. A galaxy in the constellation Ursa Major is moving away from us at about 15,000 km/sec. What is the distance to the galaxy? _____

© Mark Twain Media, Inc., Publishers

The Redshift

Astronomers have a reliable method for measuring the radial velocity of a galaxy that is moving away from us. It is called the **redshift**. Suppose that a galaxy was not moving toward or away from the earth and was emitting light. Then suppose that the galaxy started to move away from the earth and we measured the wavelengths again. We would notice that the light that the galaxy gave off would appear to have longer wavelengths. This would be true for all wavelengths of light. The faster that the galaxy moved away, the longer the wavelengths would be. The longer wavelengths are said to be **redshifted** because, for visible light, their wavelengths get closer to the wavelengths of red light.

We can write a mathematical value that relates the speed at which the galaxy is moving away to the amount of increase that we see in a particular wavelength of light. It is easier, however, if we first compute a value called the **redshift**.

$$\text{redshift} = \frac{\text{increase in wavelength}}{\text{original wavelength}}$$

After we get the redshift, we can do the following calculation to find the radial velocity of the galaxy.

radial velocity = redshift x speed of light
= redshift x 300,000 km/sec

We can use 300,000 km/sec as the speed of light.

Scientists look for a wavelength of light that they know is emitted by a certain type of atom. They know what the wavelength would be if the atoms of this type were not moving. They measure the wavelength of the light from the atoms of this type in the moving galaxy. Then they do the following:

1. Get the increase in wavelength
   increase in wavelength = measured wavelength - original wavelength
2. Calculate the redshift
   $$\text{redshift} = \frac{\text{increase in wavelength}}{\text{original wavelength}}$$
3. Find the radial velocity
   radial velocity = redshift x 300,000 km/sec

Measurements are often reported as redshifts of a certain value. For example, suppose that an astronomer knows that the wavelength of light from a certain atom is 500 nm. She measures the wavelength of light from that type of atom in a moving galaxy to be 510 nm. What is the radial velocity of the galaxy?

increase in wavelength = 510 nm - 500 nm
= 10 nm

redshift = 10 nm/500 nm   **radial velocity = 0.02 x 300,000 km/sec**
= 0.02   **= 600 km/sec**

Name _____ Date _____

Exercises—Redshifts

5. Suppose a wavelength of light is 600 nm and the measured wavelength for that light from a moving galaxy is 605 nm. What is the radial velocity of that galaxy?

_____

6. Suppose a wavelength of light is 388.9 nm and the measured wavelength for that light from a moving galaxy is 390.4 nm. What is the radial velocity of that galaxy?

_____

7. Suppose a wavelength of light is 154.9 nm and the measured wavelength for that light from a moving galaxy is 159.4 nm. What is the radial velocity of that galaxy?

_____

8. An astronomer reports a galaxy with a redshift of 0.025. What is the radial velocity of that galaxy?

_____

9. An astronomer reports a galaxy with a redshift of 0.005. What is the radial velocity of that galaxy?

_____

10. An astronomer reports a galaxy with a redshift of 0.001. What is the radial velocity of that galaxy?

_____

The Age of the Universe

We can use the Hubble constant to obtain a very rough estimate of the age of the universe. With some elementary mathematical reasoning, it can be shown that an estimate of the age of the universe is just 1 divided by the value of the Hubble constant. The result will come out in years, if the Hubble constant is converted to units of km/year per km.

For example, the Hubble constant value that we used in the first exercises of this lesson was 20 km/sec per million light years. This number converted to km/year per km is about 6.7×10^{-11} km/year per km. The age of the universe is

age = $1/6.7 \times 10^{-11}$
    = 1.5×10^{10} years
    = 15 billion years

The universe has been around for a long time. We think that our solar system was formed only about 4.5 billion years ago. It is pretty young compared to the universe.

Lesson Ten—Teacher's Page

Comments

I. The value for the Hubble constant that was used in the problems in the math section of this lesson is a representative one. It is not a precisely determined constant, however, and different values emerge from different studies. Values between 15 km/sec per million light years and 30 km/sec per million light years may be encountered. It might be interesting to calculate the age of the universe for your students using some extreme values for the Hubble constant.

    It should also be noted that the age of the universe that was calculated in the math section assumes that the universe has expanded at a uniform rate. In reality, however, it should have expanded much faster during its early existence.

II. If you generate more problems for your students to solve with the redshift equation, you should make sure that the radial velocities that result are small compared to the speed of light. If they get too close to the speed of light, they must be computed with a relativistic version of the redshift equation.

$$\text{radial velocity} = \frac{(\text{redshift} + 1)^2 - 1}{(\text{redshift} + 1)^2 + 1} \times 300{,}000 \text{ km/sec}$$

Class Activity

    You can perform a crude demonstration of the expansion of the universe with a balloon. Paint some dots on the deflated balloon to represent galaxies. Then blow up the balloon. The dots will all get farther from each other. If you were viewing the rest of the dots from a vantage point on one of them, all of the others would appear to be receding from you. Keep in mind that this is only a pseudo two-dimensional approximation of a three-dimensional expansion.

    Actually, a loaf of raisin bread that rises would be a better analogy, because it is three-dimensional. As the bread rises, all of the raisins would get farther away from one another. Unfortunately the bread loaf is too opaque and the rising is too slow.

Answers to Questions (page 122)

1. d   2. c   3. a   4. b   5. c   6. b   7. b   8. a   9. c   10. d

Answers to Math Exercises (pages 123–125)

1. 100 million light years   2. 200 million light years
3. 60 million light years   4. 750 million light years

5. ~ 2,500 km/sec   6. ~ 1,160 km/sec   7. ~ 8,700 km/sec
8. 7,500 km/sec   9. 1,500 km/sec   10. 300 km/sec